WHAT IS YOUR MIND?

EXPLORING THE MOST PROMISING SCIENTIFIC THEORIES OF CONSCIOUSNESS

Michael J. Daw

imprint-academic.com

Published in the UK by
Imprint Academic, PO Box 200, Exeter EX5 5YX, UK

Distributed in the USA by
Lightning Source,
La Vergne, TN 37086, USA

ISBN 9781788361422 Paperback

A CIP catalogue record for this book is available from the
British Library and US Library of Congress

This product complies with the General Product Safety Regulations
introduced by the European Union in 2024.

EU GPSR Authorised Representative:
LOGOS EUROPE, 9 rue Nicolas Poussin,
17000, LA ROCHELLE, France
E-mail: Contact@logoseurope.eu

ia

imprint-academic.com

Contents

Foreword by Dean Radin v

Introduction 1
Chapter 1. What is Your Mind Anyway? 4
Chapter 2. Consciousness Arises from the Brain 11
Chapter 3. Can Consciousness Exist Beyond the Brain? 23
Chapter 4. Can Consciousness Transcend Space and Time? 34
Chapter 5. Can Consciousness Directly Influence Matter? 47
Chapter 6. Do We Have Free Will? 59
Chapter 7. Consciousness is Quantum 67
Chapter 8. Consciousness is Hyperdimensional 85
Chapter 9. Consciousness is Field-Like 101
Chapter 10. Consciousness is Information 117
Chapter 11. Consciousness is All There Is 136
Chapter 12. How Effective Are These Theories? 152
Chapter 13. Towards a Better Theory of Consciousness 161
Chapter 14. What Does It All Mean? 177

Acknowledgements 190

Appendix. Philosophical Classifications for Theories of Consciousness 192

Endnotes 194

Dean Radin

Foreword

What is your mind? This is not just an academic question. How we answer it, and whether we even ask it seriously, may determine the future of our species and our planet.

After four decades conducting consciousness research at institutions including Princeton, Edinburgh, SRI International, and the Institute of Noetic Sciences, I can tell you that the evidence points toward a conclusion many find unsettling: our scientific worldview is incomplete. Perhaps woefully incomplete. Unlike the conventional neuroscience view that consciousness emerges from brain activity, there's growing evidence that it's far more surprising than that. And if that conclusion continues to be confirmed, then nearly everything about how we understand ourselves, relate to each other, and interact with the world needs to be reconsidered.

This is why Michael Daw's book arrives at such a critical moment. We face a rash of existential challenges, including environmental collapse, social fragmentation, technological disruption, and the potential emergence of artificial consciousness. While the current materialist paradigm in science has been remarkably successful, it seems ill-equipped to address these crises. We face not merely technical problems requiring better engineering, but fundamentally a crisis based on a worldview that views mind as hunks of brain matter, humans as isolated individuals, and nature as a resource to exploit.

Daw demonstrates that there is another possibility. The cross-disciplinary scientific study of consciousness has exploded in recent decades, producing hundreds of theories ranging from purely brain-based accounts to radical proposals suggesting that consciousness may be fundamental to reality itself. What we've lacked is a clear, fair-minded guide through this intellectual landscape, one that takes seriously both mainstream neuroscience and the controversial evidence that challenges it. That is exactly what this book provides.

Daw's treatment of 'weird phenomena'—including out-of-body experiences, telepathy, precognition, and psychokinesis—is crucial. Many scientists have either been too timid to wade into these controversial topics, or they dismiss them without examination. But the evidence deserves serious consideration, not only because weird phenomena have been reported throughout history and in all cultures, but because decades of controlled experiments from laboratories worldwide suggest that consciousness may in fact transcend our skulls. If even a fraction of the lab evidence withstands scrutiny—which in my opinion it clearly does—then it means one or more core assumptions in science are fundamentally incomplete. And that matters because worldviews have consequences.

A naïvely materialist framework, one based largely on a nineteenth century understanding of physical laws, which is what the majority of scientists have learned to take for granted, reduces consciousness to brain activity. This naturally leads to certain attitudes: that mental illness is a chemical problem, that death is absolute annihilation, that other beings are essentially mechanical, that meaning and purpose are illusions. This nihilistic worldview shapes everything from healthcare to environmental policy to how we design and think about artificial intelligence. If the underlying assumptions are wrong, then we've been building our civilization on a house of cards.

Daw doesn't advocate any particular theory. Instead, he presents the evidence fairly, acknowledges sceptical critiques, and avoids both uncritical acceptance and dismissive rejection. He recognizes that no current theory, whether materialist or beyond, is complete. But he also makes clear that the question cannot be settled by philosophical arguments alone. We need meticulous experiments to help discern which of the many competing theories are closer to the 'truth', as best as we can tell.

The middle chapters explore theories that propose consciousness extends beyond the brain. This includes models based on quantum mechanics, higher dimensions, field theories, information-based approaches, and idealist philosophies. Each of these frameworks suggests different possibilities for human experience and potential. If consciousness is quantum in nature, and the quantum observer effect includes a mental component, it might explain why consciousness affects reality. If it's field-like, it might account for why we sometimes know things we shouldn't be able to know. If consciousness is fundamental rather than derivative, then we are not accidents in an indifferent universe but, as physicist John Wheeler put it, we play a key role in a participatory universe.

These questions have practical urgency. We're developing artificial intelligence without understanding natural intelligence, treating mental illness without understanding the full capacities of the mind, and facing environmental catastrophe while viewing nature as an unconscious mechanism. Building a technological civilization on a foundation of assumptions without realizing that some of those assumptions are probably wrong is a recipe for disaster. But with improved assumptions, based on solid empirical evidence, we may gain confidence that consciousness can indeed access information non-locally and directly influence matter. Then the possibility arises to develop entirely new approaches to healing, education, communication, and human development. Disaster avoided.

Who will benefit from reading this book? Scientists will find a rigorous treatment of competing theories of consciousness, with an emphasis on testability. Students will find an accessible guide through a labyrinth of complex ideas. And anyone who is personally grappling with questions of meaning, purpose, human potential, and the health of our planet will find a framework for thinking clearly about consciousness and why it matters.

Daw offers no final answers; at this point no one can. But he does provide a clear-eyed assessment of where we stand and what the evidence suggests. If you think you have the nature of consciousness already figured out, or you don't understand why anyone would be interested in studying it, this book will challenge you. If you're uncertain and seeking clarity, it will guide you. If you care about humanity's future, it will help you think more clearly about the most fundamental questions we can ask.

In sum, the question 'What is your mind?' is not just abstract philosophy. It's the question that determines who we are and what we can become.

Dean Radin, PhD

Chief Scientist, Institute of Noetic Sciences
Author of The Conscious Universe, Entangled Minds, Supernormal, Real Magic, *and* The Science of Magic

December 2025

For Nick

Introduction

Every year since 1994 in either the Arizona desert or someplace else in the world, almost a thousand interested individuals come together to discuss the latest ideas and research that aim to resolve the age-old question of 'what is your mind'. According to the website of this conference, among their number are "philosophers, psychologists, artificial intelligence proponents, physicists, meditators, altered-state researchers, [and] mystics".[1] The mystery of consciousness is, the organisers suggest, "one of science's last great frontiers".[2]

My quest to understand what is mind has been unfolding ever since I can remember. From reading in my childhood about ghosts, UFOs and spontaneous combustion, to unsuccessful attempts to bend spoons like Uri Geller, to learning meditation, to following the work of inspirational scientists of consciousness, to now (thrillingly) conducting consciousness-related research myself as part of an international community of parapsychologists.

I can't claim to understand consciousness. But then no one else can either, at least on an intellectual level. (With apologies for the spoiler.) Indeed, it's likely to be a long time before we unlock this mystery. Or perhaps a satisfying explanation is just around the corner. Some people believe the nature of the problem means that this might never happen. Regardless, the scientific field of consciousness studies is replete with mind-bending ideas of many different persuasions put forward by researchers working across diverse academic disciplines. This book examines fourteen such theories from those that suggest the mind is entirely a product of the brain to those proposing that consciousness is more fundamental to reality than our supposedly solid physical world.

Here's how the book is structured. Chapter 1 provides a brief history of enquiries into the nature of mind and offers a framework of consciousness in terms of some of its most salient properties, which, I suggest, any theory worth its salt should aim to accommodate.

Chapter 2 is a review of four groups of theories that suggest that consciousness arises entirely from brain-based processes. These are neuroscientific theories that pop up most often in discussions of consciousness in scientific publications and are known as materialist or physicalist theories because they assume that reality consists only of physical matter and forces and therefore consciousness must be entirely a product of the brain.

The next four chapters present the evidence for what I call weird phenomena. 'Weird' because, at face value at least, these phenomena contradict the premise of brain-based theories in that they seem to show consciousness persisting and exerting influence beyond the brain. We'll look at cases where consciousness seems able to travel beyond the body and migrate to another body after death (Chapter 3), communicate between minds and know the future (Chapter 4), and exert direct effects on the physical world and living organisms (Chapter 5). Along the way, we'll examine the case made by sceptics who argue that such weird phenomena can be fully explained in essentially mundane and non-weird ways. We'll address the problem of free will in Chapter 6, which, if genuine, also seems to contradict the basis of brain-based theories and hence might also be considered weird to those who hold such a viewpoint. If any of these weird phenomena are shown to be real, then this would seem to deal a fatal blow to the premise of theories in which consciousness is equivalent to processes in the brain.

The next five chapters review ten theories where consciousness is proposed to extend beyond the brain and physical body. In other words, where consciousness is non-local, that is, where it is not confined to a single place in time and space as in brain-based theories. These theories come in various flavours so we'll deal in turn with those that suggest that consciousness is related to quantum mechanics (Chapter 7), that it is to be found in higher dimensions (Chapter 8), that it is field-like (Chapter 9), that it is essentially information (Chapter 10), or that it is all there is (Chapter 11). At least in principle, these theories of non-local consciousness should be more suited to accommodating weird phenomena than brain-based theories.

In Chapter 12, I summarise how well the theories of consciousness considered in the book account for both weird and non-weird features of consciousness and in terms of being good scientific theories. In Chapter 13, I make some suggestions for how this field might make meaningful progress. This includes a proposal to integrate theories from diverse perspectives—both brain-based and non-local—which might help to advance an explanation for consciousness that is more comprehensive and powerful than any single theory alone can achieve, despite how these different approaches appear at first glance to be incompatible. I also suggest that more robust and rigorous testing of the theories is required, particularly for theories of non-local consciousness and with respect to weird phenomena, to help differentiate and refine theories of consciousness to tease out which aspects work and which don't. In the final Chapter 14, I offer a personal reflection on the importance of understanding the fundamental nature of consciousness to our future and that of our world.

I hope you enjoy this journey through theories of your mind as much as I have enjoyed writing it.

Chapter 1

What is Your Mind Anyway?

What exactly is your mind? Where does it come from? Why do we have a mind at all?

Such questions on the nature of consciousness were likely a source of wonder for some of the very first humans and they remain unanswered. One reason why they are so hard to resolve might be because of the sheer ubiquity of our experience of mind and consciousness.[i] Every activity, feeling, thought and emotion we've ever had, are having right now, or will ever have, takes place within an ever-present consciousness. Falling in love, being ill, watching a sunset, reading, going for a swim, indeed, every experience we can possibly imagine—including imagination itself—is a manifestation of consciousness. Consciousness is both experience itself and the medium in which experience takes place. Consciousness is not separate from experience, but when any experience ends, consciousness remains.

The prevailing view among much of the scientific community is that our consciousness comes to a permanent end when our brain dies, although it might disappear for a while in certain circumstances, for example, when we're given a general anaesthetic. However, an increasing number of scientists suggest that our minds may continue to exist even after our bodies have long decayed. The question of whether consciousness requires a brain and a body to persist is an

i The words 'mind' and 'consciousness' are subtly different in meaning. We probably associate the word mind more as an individual personality and consciousness as more of an underlying substrate. But then people also talk of a 'collective mind' so the distinction is not clearcut. This debate is not central to this book and I use the terms more or less interchangeably.

important point of distinction—and indeed contention—among competing explanations for consciousness.

The oldest writings on the nature of consciousness may be the ancient Indian texts known as the *Vedas*, which date back to the fifteenth century BCE.[3] The *Vedas* represent a complete metaphysics with descriptions of gods and rituals said to help achieve freedom from the cycle of birth, death and rebirth.[4] In China, ideas about the nature of consciousness can be traced to the eleventh century BCE and include concepts such as the Tao, a term meaning the entirety of all there is and which has two essential complementary qualities, yin and yang.[5] In ancient Greece, figures such as Pythagoras, Socrates, Plato and Aristotle were engaging in heated debates about consciousness from the sixth century BCE. Modern philosophy is generally considered to date back to the seventeenth century with the writings of René Descartes. Descartes suggested that reality is made up of two fundamentally separate and different kinds of substances—one being physical matter that we can see and feel, and the other being mind, non-material thought-stuff—a viewpoint that came to be known as dualism.[6]

Much more recently, in just the last few decades, questions regarding the nature of consciousness have been taken up within the domain of science. This is despite ideas of mind and consciousness previously being considered as beyond empirical study and amenable only to philosophical enquiry. In 2018, a special joint supplement to the journals *Nature* and *Scientific American*—two of the biggest names in science publishing—proposed what they considered to be the six 'biggest questions in science'.[7] Number three—sandwiched between 'what is dark matter?' and 'how did life begin?'—was the quest to develop a scientific theory of consciousness.

This book is about the current state of this big question. I describe some of the most promising and compelling theories of consciousness that have been proposed by scientists working in fields such as neuroscience, biology, physics and psychology. It is not the intention of this book to be comprehensive. Indeed this would be challenging as one recent paper suggested that there are over two hundred scientific theories of consciousness.[8] Such an approach would be exhausting and perhaps not so illuminating. My aim instead is to present some of the theories that might be considered to be the most interesting and influential, along with why they fall short. (Because no theory of consciousness is yet considered final and comprehensive.) I also present some ideas for how we could make more significant progress in this field. A word of caution before we begin. Given that questions about the nature of consciousness have been with us for millennia, a little humility is due. As Susan Blackmore and Emily Troscianko wrote at the start of their book summarising consciousness: "if you think you have a solution… you haven't understood the problem."[9]

But how might we define consciousness? What is it that any worthwhile theory should be trying to explain? Unfortunately, the field of consciousness studies is often criticised for its vagueness in defining exactly what consciousness is[10] and so this chapter is an attempt to provide a working framework for this question. I suggest three important and distinctive features of consciousness that we might reasonably expect any effective theory of consciousness to address. These can act as a starting point that we can use to assess the incredibly diverse range of theories of consciousness that have been proposed, in particular those that are discussed in this book. These defining features are that:

1. Consciousness is fundamentally subjective, private and personal, and encompasses rich and deep experiences with qualities that are hard—maybe even impossible—to describe or quantify precisely.
2. There are a number of distinct states of consciousness, such as waking and sleeping, each with its own unique features.
3. There is something about animate and conscious beings that is fundamentally different from things that are inanimate and seemingly not conscious. This question is particularly pertinent to current discussions about whether artificial intelligence (AI) can be conscious.

Let's examine each of these in more depth.

Rich Personal Experience

We experience our lives as a rich tapestry of awareness through our interactions with the world and with each other. We embody our own unique blend of opinions, desires, emotions and feelings. We have what we might describe as subjective awareness. I know what it feels like to be me, you know what it feels like to be you and, as the philosopher Thomas Nagel declared, there is something it is like to be a bat.[11] We experience thoughts as private and personal, and we believe ourselves to be a continuation of the same person we have always been despite inhabiting a body that has changed over time. We also have rich immersive experiences of the world in terms of sights, sounds, feelings, tastes and smells that have meaning for us.

Consciousness seems apart from any other phenomenon that science attempts to explain in that it is only directly accessible to the conscious person themselves. All other phenomena of the world—such as temperature, mass, velocity and so on—are in principle accessible from a third-person perspective and can therefore be tackled objectively (although, as we'll see later in this book, this idea—long taken for granted—has recently become rather more open to question). In addition, constructing physical descriptions of rich personal experiences, which are known to those who study consciousness as 'qualia',[12] seems an entirely different matter from what it actually feels like to have these experiences. David Chalmers calls bridging this explanatory gap the 'hard problem of consciousness'.[13] Hard, of course, being a relative term when compared to the so-called easy problems of consciousness of identifying which neurons or regions in the brain light up when we have particular experiences, otherwise known as the 'neural correlates of consciousness'.

Frank Jackson has been particularly eloquent in his description of the hard problem of consciousness and offers some vivid examples of qualia. He writes:

> Tell me everything physical there is to tell about what is going on in a living brain, the kind of states, their functional role, their relation to what goes on at other times and in other brains, and so on and so forth, and be I as clever as can be in fitting it all together, you won't have told me about the hurtfulness of pains, the itchiness of itches, pangs of jealousy, or about the characteristic experience of tasting a lemon, smelling a rose, hearing a loud noise or seeing the sky.[14]

A number of other philosophers have also pointed out the difficulty of describing consciousness and experiences in physical terms when they are essentially entirely subjective. These include Bertrand Russell, who wrote about the direct subjective awareness of conscious experience, what he called 'knowledge by acquaintance',[15] and Ned Block, who highlighted the difference between what we are able to describe to another person about an experience we have had compared with the felt full and rich depth of this experience, which defies complete description.[16] Whilst this issue is indeed, in Chalmers' understated description, a hard problem, it is nevertheless one that theories of consciousness should attempt to address.

States of Consciousness

The second key feature concerns states of consciousness. Our subjective felt consciousness (our mind) routinely transitions between recognisably distinct states, each with its own characteristics and associated brain patterns. The one you and I currently occupy as I write this sentence and you read it is the waking state of consciousness. At some point later today (if insomnia doesn't strike), we'll experience two other states of consciousness—deep sleep and dreaming. We also experience brief transition states of consciousness between sleeping and waking—hypnagogia (when we're falling asleep) and hypnopompia (when we're waking up).

Beyond these mundane or natural states of consciousness that everyone regularly experiences, there are also more exotic states that occur less frequently. These include hypnosis, most often used for therapy or for public entertainment, and which can be self-induced or brought on by someone else, and transcendental or mystical states associated with spirituality or religion accessed through practices such as shamanic rituals or meditation. Psychedelic states of consciousness arise from ingesting drugs such as LSD, a chemical synthesised in the lab; psilocybin, magic mushrooms; and mescaline, derived from cacti. These and other substances are capable of inducing sometimes dramatically-altered perceptions of time and space and a variety of other wild experiences usually categorised in the academic literature as hallucinations, but which may hold deep meaning (and reality) for those who experience them. All of these states are known as evoked states of consciousness because they do not in the main occur spontaneously and usually require some kind of deliberate intervention.[17] There are also states of consciousness associated with brain damage and injury or conditions such as dementia. The question is, how might theories of consciousness account for this wide variety of different states that we might experience?

Determining Between Conscious and Non-Conscious Entities

Our third feature relates to whether we can reliably distinguish between entities that are conscious and those that are not.[18] What is it about beings that we assume to be conscious, such as humans, cats and octopuses, that differentiate them from things we assume are not, such as tables, waterfalls and atoms? Douglas Hofstadter and Daniel

Dennett point out that we notice a correspondence between our inner experience and our behaviour such as when we experience a sharp pain from touching a hot surface, yell and sharply withdraw our hand, and therefore suggest that similar behaviour in others indicates that they too are having a similar private experience of pain. In other words that they are conscious just like us.[19]

But such judgements are highly subjective. People's willingness to make an attribution of consciousness can be affected by superficial similarities, so that it might be easier to ascribe an inner world to animals that look a bit like us such as gorillas and chimpanzees than it might be to, say, slugs and starfish. Making demarcations on such arbitrary terms is clearly unsatisfactory, so a theory of consciousness would ideally provide explicit criteria for deciding which entities have consciousness and which do not. Alternatively, a theory might propose that all entities are conscious, but this should be without rendering the notion of consciousness so uniform and ubiquitous that the term loses any explanatory value.

The question of how to determine whether any particular entity is conscious or otherwise is becoming increasingly salient due to recent dramatic advances in AI and the rapid spread of this technology across many aspects of modern life. In 2022, Blake Lemoine, a computing engineer working for Google, claimed that one of the firm's AI systems designed for free-flowing conversation, known as Lamda, had become sentient and therefore should be entitled to certain rights.[20] Google subsequently dismissed these claims and fired Lemoine.[21] However, AI chatbots can seem disarmingly real to the extent that many people interact with them as if they are conscious beings, for example by saying please and thank you,[22] although experts dismiss such notions as excessive anthropomorphism.[23] What might our theories of consciousness have to say about how to determine whether any entity is conscious or non-conscious and the issue of AI in particular?

***

These three issues related to consciousness are, of course, only one approach to frame the problem. Others may suggest different emphases. However, resolving or accounting for these and other defining issues should be central to any satisfactory theory of consciousness. There are, I argue, other striking features of consciousness that any theory aiming to be comprehensive should address. As we shall see, there is an accumulating body of evidence that suggests that consciousness appears to express characteristics beyond the basic defining properties outlined in this chapter. Characteristics that appear to show that consciousness may be able to reach beyond the brain to perceive and share information, and to affect the physical world in ways that seem to challenge our current understanding of reality. But not everyone is comfortable with accepting such 'weird' phenomena. So let us for now defer discussion of these questions and turn our immediate attention to what are currently seen as the leading scientific theories of consciousness which assert that no such awkward phenomena exist. These are theories where consciousness is explained as arising entirely from brain processes.

Chapter 2

Consciousness Arises from the Brain

In the last chapter, I proposed a view of consciousness based on three of its defining features with the aim of providing a benchmark with which to assess the scientific theories included in this book. Now is the time to begin to examine these theories in more detail and use this framework to consider how successfully they might be able to explain the nature of consciousness.

The first four theories we'll look at—or, more specifically, the first four *groups* of theories—all take as a starting point the assumption that the universe is made exclusively from physical material and fields, the stuff of physics, chemistry and biology. Hence the only possible explanation for the mind's rich tapestry of thoughts, imagination, feelings and experiences is that it must arise from something we can touch or measure. The prime candidate for this physical source of consciousness is the brain.

The study of the brain reaches back to ancient times but innovations in sophisticated imaging techniques over the last few decades have revolutionised our understanding of this highly complex organ. Successes in diagnosing various brain-based diseases and in identifying some of what are called the 'neural correlates of consciousness'—those parts of the brain that activate in conjunction with different experiences—have led some neuroscientists to become confident that an explanation for how the brain produces consciousness is only a matter of time.

Francis Crick, the co-discoverer of DNA, summed up this brain-based viewpoint in what he called the 'astonishing hypothesis'. He stated that "your joys and sorrows, your memories and your ambitions, your sense of personal identity and free will, are in fact no more than the behaviours of a vast assembly of nerve cells and their associated molecules". And so, he concluded, "you're nothing but a pack of neurons".[24]

This approach would have it that, essentially, the brain, or at least certain brain processes, is equivalent to the mind. In this chapter, we'll summarise the most prominent brain-based theories, which largely fall into four main approaches.[25,26] Three of these, known as higher-order, global workspace and recurrent processing theories, all propose that most of the activity of the brain is unconscious but that what we experience as consciousness is equivalent to its high-level processes. Predictive processing theories take a slightly different tack in suggesting that the brain works as a kind of prediction machine, continually testing its best guesses about the nature and content of the outside world against incoming data from the senses, and then refining its model of reality accordingly.

Let's look more closely at these four types of brain-based theory.

Higher-Order Theories

Higher-order theories propose that our consciousness is made up of high-level mental states that are representations of more primitive and basic mental states.[27] Basic mental states are essentially raw perceptions of the outside world happening within the brain. For example, an image of an apple might be in our line of sight but we might not necessarily be aware of it. Some portion of the brain will be engaged in processing this image but it remains an unconscious perception unless and until, say, we feel hungry, at which point we might notice it. This is an example of a basic mental state (the unconscious perception of an image) entering a higher-order state. In this case, this higher-order state is our experience of consciously noticing the apple.

All sorts of evidence have been presented to suggest that this is how our minds work, with the brain engaging in multiple layers of conscious and unconscious processing.[28] One example of this is the experience we may have had of driving somewhere only to realise when we arrive that we have done so without any conscious thought whatsoever, perhaps because we've been engaged in deep conversation with someone else in the car. Yet we've arrived perfectly safely having navigated all sorts of potential obstacles, such as other cars, pedestrians, traffic lights and so on. The process of driving has become so engrained that it can be done automatically—at lower orders of unconscious perception—whilst our higher-order conscious awareness is focused on something else.[29]

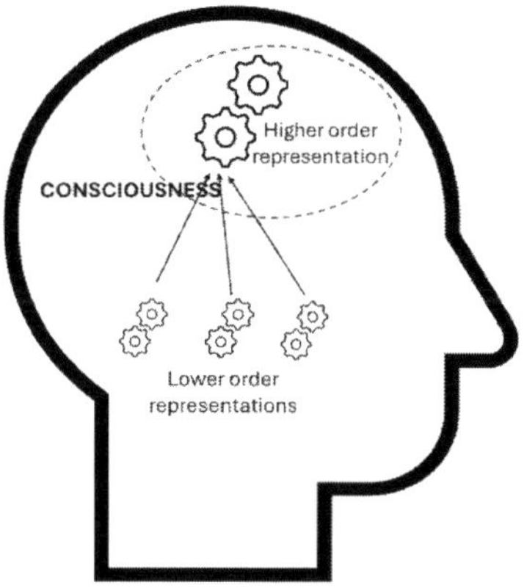

Figure 1. Higher-order theory of consciousness

Global Workspace Theories

Our second group of theories consists of global workspace theories. These suggest that background thoughts in the form of brain processes enter our awareness when they are 'broadcast' across a shared workspace that is distributed across the brain. This broadcast is essentially what we feel to be consciousness. Certain experiences that we know definitively involve conscious awareness, such as accessing memories, paying attention to something, evaluating a problem and speaking all involve mental states that are available to the whole brain. Hence the idea of a global workspace.

The organisation of consciousness in the brain as a global workspace is somewhat similar to how some forms of AI work. AI usually involves a centralised global workspace that is shared and accessible by all the separate individual computational processes that contribute to the whole system. Hence, proponents of these theories are suggesting that the architecture of the brain works like a computer program and that consciousness is essentially a 'computational property'.[30] According to this theory of consciousness, consciousness could therefore arise within machines, at least in principle.

The premise of global workspace theories is that the brain works in a decentralised fashion with processes occurring in many different regions.[31] However, the frontal and prefrontal cortex of the brain are considered to be the hub of the global workspace with a dynamic network of connected areas spread across the brain.[32,33] Global workspace theories recognise that a great deal of sophisticated cognitive processing takes place without any conscious awareness, but that being conscious of certain stimuli when they enter the workspace enables additional and useful cognitive processes to occur.[34] If so, this would explain why consciousness has evolved as a feature of complex brains.

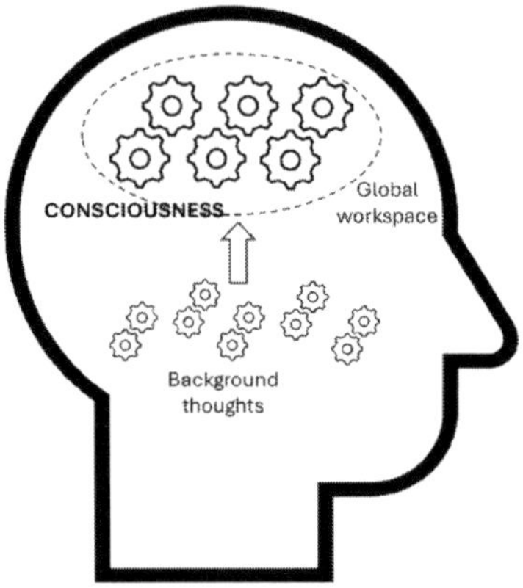

Figure 2. Global workspace theory of consciousness

Recurrent Processing Theory

Like higher-order and global workspace theories, recurrent processing theory also regards consciousness as being essentially equivalent to high-level brain functions. This theory emerged particularly from empirical observations of how the brain manages visual information. Studies and experiments examining this process seem to show that visual perception involves four distinct stages.[35]

The first of these stages is the brain determining from the visual system various features of whichever object happens to be in our line of sight—such as its colour, shape, motion and so on. This is unconscious. The second stage involves initial and also unconscious bodily reactions to the object such as eye movements or limb movements. Flinching or catching a fast-moving ball might be examples of unconscious reactions that happen in this stage.

The third stage is where consciousness enters the picture. This is the experience of consciously seeing something in which our brain extracts more features about the object. This stage involves a back-and-forth between neurons in the visual cortex. This is the recurrent interaction of neurons that gives this theory its name. We are conscious within this stage but only at a relatively low level of attention. The fourth and final stage is where we are fully engaged with what we are seeing. This involves recurrent processes across the whole brain. This is a kind of 'knowing' where what we're seeing might trigger recognition or memories or other thoughts, or it might provoke some internal dialogue or cause us to talk about it with others. We might make decisions related to what we're seeing.

Essentially, this theory suggests that the first and second stages do not involve recurrent processing and so are unconscious; these stages can be observed in anaesthetised animals who are definitively not conscious. The third and fourth stages of brain processes involve recurrent—back-and-forth—neuronal activity and it is these that equate to consciousness. These 'seeing' and 'knowing' stages are accompanied by us being aware—of being conscious—of whatever it is that we're perceiving.

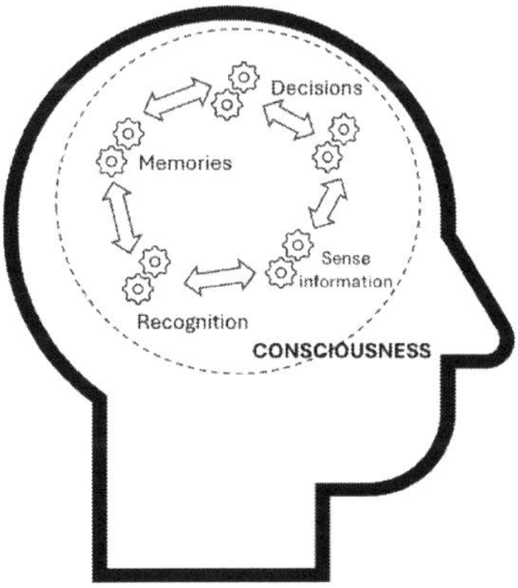

Figure 3. Recurrent processing theory of consciousness

Predictive Processing Theories

The first three groups of brain-based theories we've looked at in this chapter—higher-order, global workspace and recurrent processing theories—all suggest that consciousness is equivalent to some higher level of brain processing. Predictive processing theories offer a somewhat different model for consciousness, suggesting that our brains work like prediction machines, continually making best guesses about the way the world is and then refining and updating these models according to data received through the senses. These theories emphasise the role of top-down processing by the brain in shaping and enabling our conscious perceptions—the brain quite literally determines what we see, hear, feel, touch, smell.[36]

Proponents of predictive processing theories suggest that consciousness is the interaction between feed-forward (prediction) and feed-back (sensory input) mechanisms, which places this theory within the field of cybernetics. (Although it is fair to point out that some theorists don't make the claim that these processes actually *equate* to consciousness, rather that these processes are merely *involved* in aspects of consciousness.) For those of a mathematical bent, this theory is sometimes compared to Bayesian statistics, which works by calculating a probability of the way the world is based on an initial starting point that is taken from personal experience, prior beliefs or previous experiments, which is then updated when new data is received.

Whilst it goes against how we might intuitively think our senses and brain interact with each other—we look, then we process what we see—this emphasis on perceptual experience being modified by predictions from the brain can be demonstrated with optical illusions such as the familiar picture of a silhouette that looks either like a vase or a side view of two faces where the brain flip-flops between alternative interpretations. It is us (meaning our brain) that determines what we see despite the image not actually changing. Much current neuroscience research focuses on demonstrating how our consciousness plays an active and creative role in forming impressions of what we perceive. The findings from many of these experiments show us that we are not the objective and neutral observers of reality we might like to think we are. What we expect determines what we see. Neuroscientist Anil Seth suggests that perception is a 'generative act' in which our expectations help shape the brain's model of the world. Such expectations could be related to the way we think, our personal preferences and the particular social groups, society and culture in which we live. In other words, the way we are determines to a significant extent what it is that we actually perceive.[37]

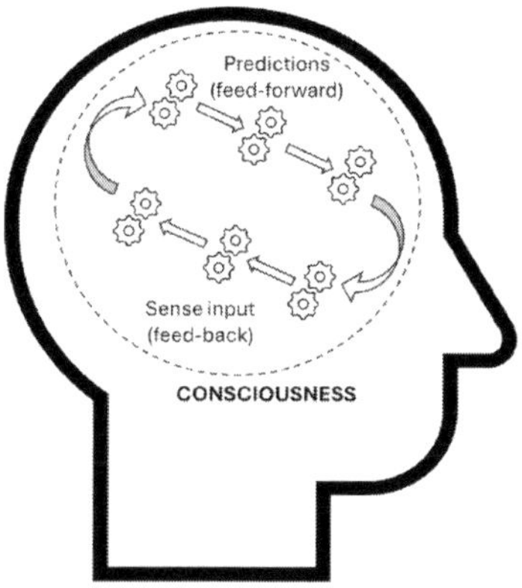

Figure 4. Predictive processing theory of consciousness

Features of Consciousness

These are the four most prominent approaches from neuroscience for how the brain might produce consciousness. In the last chapter, we outlined three defining features of consciousness that any good theory should be able to accommodate. So how might these brain-based theories fare against these criteria? Let's examine each in turn.

Rich Personal Experience

The first defining feature of consciousness concerns how consciousness feels personal and subjective and is imbued with rich experiences (qualia). That consciousness is personal and private and feels continuous throughout our lives seems relatively straightforward to account for from a brain-based perspective in which what we think of as our self is essentially just our brain (we're just 'a pack of neurons', as Francis Crick would have it). Therefore, given that the brain is encased in the skull, it makes sense that the feeling of mind it produces is private and personal and not shared. That we feel like the same person we always have been can be explained by how the brain evolves very gradually. Neurons are created and die continuously throughout our lives. But it's a slow process—neurons are the longest living cells in the body,[38] which means that the overall structure of the brain remains relatively static over time. Therefore, it makes sense that it feels like we are always the same person, whatever our age.

In terms of qualia, the first three approaches we looked at—higher-order, global workspace and recurrent processing theories—all propose that it is the various configurations of high-level processes within the brain that are equivalent to subjective experiences. Whether these take the form of high-level representations (higher-order theories), broadcasts to a distributed workspace (global workspace theories) or back-and-forth processes between neurons (recurrent processing theory), each of these is in some way equivalent to conscious experience. Low-level processes on the other hand are, by definition, unconscious and so do not encompass subjective qualities.

Predictive processing accounts for qualia somewhat differently through the idea that it is the predictions our brain makes that equate to subjective experiences. Whilst we focused in our description of this theory on predictions and perceptions of the outside world, our brain is also making predictions and perceptions of what's going on with our bodies, for example, if we are shivering. This means that there are two parallel processes going on within the brain—predictions about the external world and predictions related to the state of our bodies, both of which are being continuously updated, the first by information our senses get from the environment and the second by bodily feelings. This combination of external and internal perceptions helps to determine what we need and want. Therefore, so the argument goes, the subjective qualities we perceive in terms of any experience are to be found in how they make us feel and act, which comes from a combination of the two streams of prediction-perception loops. For example, the subjective feeling of being cold comes from perceiving a change in our bodily state (shivering or getting goosebumps) and relating this to a change we perceive in the outside world (someone's left the door open). This is how and why our predicted models of the world—both external and internal—are equivalent to our conscious experiences, or qualia.[39,40]

From these perspectives, qualia can all be thought of in terms of some form of integrated neuronal activity. Whilst each of these theories emphasises a different aspect of how the brain works, they share a view that consciousness either stems from or is equivalent to some coordinated activity of neurons in the brain. But why and how is it that brain processes have the subjective nature they do? Are they sufficient to explain the rich experiences of consciousness, such as feelings of awe, beauty and pain? According to this paradigm, we are biological machines and there is little attention paid to rich subjective experience. To quote once again the words of Francis Crick, such a view is indeed an 'astonishing hypothesis'.[41] Unfortunately, it is difficult for brain-based approaches to explain and differentiate between the unique felt qualities of completely distinct kinds of experiences, such as the taste of sugar and the sound of a bell. In other words, the hard problem of consciousness—explaining the gap between the nuts and bolts of physical processes and our felt experiences—is left hanging.[42,43,44] It is almost as if the central question of consciousness that should be the core of any comprehensive theory is somehow dodged as altogether too awkward.

States of Consciousness

The second of our defining features concerns states of consciousness. How might brain-based theories accommodate the distinct qualities associated with different states, such as those encountered when dreaming or in deep sleep or in meditation? Because these theories focus on what is happening in the brain, they might be expected to show more promise here. In principle, they should be able to account for different states of consciousness with reference to distinct configurations of the features that each proposes as being equivalent to consciousness.

Brain-based theories of consciousness focus on the waking state of consciousness and so explanations for other states of consciousness tend to be couched in terms of how brain processes in other states might deviate from this waking baseline. Higher-order theories propose that consciousness equates to high-level representations of unconscious processes. Throughout the day these take the form of representations of basic perceptions. However, when we are dreaming, these basic perceptions are absent or much weaker so it might be that in this state our brains manifest free-floating high-level representations in the form of dream characters and objects that have no corresponding lower-order counterparts. In deep sleep or under general anaesthetic there are no higher-order representations, and therefore no conscious awareness.[45]

Some proponents of global workspace theories suggest that how different states of consciousness correspond to differences in brain activity provides supporting evidence for this perspective. For example, studies have shown that whilst low-level neuronal activity occurs in both dreaming and deep sleep, it is only when dreaming that there is high-frequency activity in the areas associated with the global workspace. This accords with an expectation that there will be some similarities in brain activity between dreaming and the waking state but both will be different to deep sleep. Furthermore, in the lucid dreaming state—which is when a person knows they are dreaming—there is increased activity in the areas associated with the global workspace compared with non-lucid dreaming, which again is what we would expect if this model is correct.[46]

As with higher-order and global workspace theories, recurrent processing theory also suggests that states of consciousness are distinguished in terms of the degree of high-level brain processes that occur in any particular state. The theory's main proponent, Victor Lamme, suggests that different states represent 'non-linear gradations' of consciousness, with sleep and anaesthesia, for example, being examples of 'unconscious' states.[47] In other words, there is a scale of consciousness with wakefulness having most awareness (most recurrent processing) and other states of consciousness having lower levels of awareness (less recurrent processing).

For predictive processing theories, the suggestion is that each state of consciousness is associated with a particular relationship between the brain's predictions and information from the senses. For example, when we're asleep there is minimal sensory input, which means that, whilst dreaming, the brain can make predictions about reality that are unencumbered by corrections from the physical world, leading to what can be wild and vivid dreams. Other states are defined by other types of interactions between the brain's predictions and what the body's senses are telling it. For example, states of consciousness associated with using psychedelic drugs manifest in the often highly exotic experiences people report whilst tripping.[48,49,50]

However, most of the focus of research regarding brain-based theories has so far been on accounting for a basic presence or absence of consciousness according to the principles of the theory in question rather than explaining any more varied states of consciousness. Hence, global workspace theory pays attention to whether the global workspace that is suggested to be the root of consciousness remains intact and operational when anaesthetic (or some other factor) causes an absence of consciousness. Other brain-based theories have similar proposals relevant to their particular high-level process of interest. More sophisticated research in the context of these theories regarding more diverse states of consciousness is still in its infancy or yet to be considered.[51]

Determining Between Conscious and Non-Conscious Entities

The third feature of consciousness concerns how theories might distinguish between conscious and non-conscious entities. Which pointers might be provided by these brain-based theories to determine whether one entity has consciousness and another does not?

Each of these theories offers a similar response to this question. This is that a conscious entity is defined as one that can execute the processes they regard as equivalent to consciousness and a non-conscious entity is one that can't. For example, for higher-order theories, the boundary of consciousness lies at a threshold of complexity where a system can exhibit two levels of information processing—high and low.[52] One might suggest that an amoeba, having only a single cell, is too simple to have two orders of cognitive processing and therefore does not have consciousness, whereas an octopus, with its complex and intricate nervous system, might have such a capacity for dual levels of brain processing. Global workspace theories suggest that, in order for any entity to be conscious, it requires a brain or something similar that has sufficient complexity to involve a global workspace model.[53] Whilst that would obviously rule out inanimate objects such as a table or a light bulb, it would not necessarily preclude other animals having consciousness through something like a global workspace operating in their cognitive systems, even if these are very different to ours.[54,55] One could make very similar arguments for recurrent processing and predictive processing theories regarding their criteria for consciousness.[56]

This approach means that it only takes a relatively small set of mechanisms to enact any of the functionality suggested by these theories as being equivalent to consciousness. The problem with this is that it is possible to conceive of relatively simple structures that embody these processes that are probably not conscious. Indeed, global workspace theory is modelled to some extent on how some relatively basic AI computer systems function, but most theorists would probably not conclude that these AIs are necessarily conscious beings. Unfortunately, brain-based theories do not seem to offer any clear threshold of complexity that we can apply to determine when particular systems can be said to be conscious or not, meaning it all seems to come down to a matter of degree.

Perhaps then we should class all animals, and maybe even plants and some computers and machines, as having consciousness? Victor Lamme has pointed out that just because a wide range of entities may have consciousness, this does not necessarily mean that the form of their consciousness is the same or similar to the rich depths of consciousness that we experience. Perhaps we can say instead that other entities are indeed conscious but that their experience is something much more basic, perhaps more akin to states within our minds that we only barely perceive.[57]

***

This chapter is a brief summary of the four most prominent theories suggesting that consciousness is either equivalent to, or arises from, brain processes. We've looked at some of the issues that these theories have in accounting for what we might call those basic features of consciousness, namely how our minds are personal and encompass rich subjective experiences, how we experience distinct states of consciousness and what it is that differentiates conscious entities from those that do not have consciousness. But what if there was convincing evidence of consciousness-related phenomena that brain-based theories could not possibly account for, even in principle? Evidence that consciousness might not need a functioning brain, that it is not confined to the walls of the skull, that it transcends space and time and could directly influence matter? Is it possible that we are more than mere biological robots slave to the deterministic machinations of our grey matter? Such evidence would surely sound the death knell for any proposal that consciousness can somehow arise from the brain. Over the course of the next four chapters, we will examine the case both for and against the reality of phenomena such as these.

Chapter 3

Can Consciousness Exist Beyond the Brain?

In the last chapter, we discussed theories proposing that brain activity produces consciousness. The implication of these theories is that consciousness is localised within the brain and without the physical body there can be no mind and no consciousness. This idea underpins almost all of the discussions about consciousness that appear in the scientific press. However, whilst pervasive, this view remains only an assumption with as yet far from conclusive supporting evidence. Indeed, there are certain phenomena that have been reported throughout human history that threaten the very foundations of this paradigm. If they are as they appear to be then such experiences mean that consciousness could function and persist beyond the brain, transcend space and time and influence physical matter directly. Such weird phenomena are often also known as 'paranormal' phenomena. Another issue for brain-based theories is that we appear to exhibit free will, which is in accordance with commonsense but is contrary to the view of many brain-based theorists who suggest that our choices are illusory and instead just a product of deterministic brain activity. If any of these phenomena were shown to have any basis in reality then the notion that consciousness arises solely from the brain would be severely undermined. These weird phenomena are the focus for this and the following three chapters.

But why call such phenomena 'weird'? Many such experiences have been reported by significant numbers of people so could be considered quite normal. Indeed, a belief in free will is almost ubiquitous and surveys suggest that around 40% of the population have had paranormal experiences.[58] However, weird phenomena point to a universe that is not mechanistic. Their very existence seems to threaten the orthodox scientific paradigm that has been dominant since the Enlightenment. Hence in this respect these are weird phenomena. But there's another reason for using this word. The linguistic root of 'weird' is 'wyrd' and this refers to the ancient Anglo-Saxon concept of fate or destiny, typified perhaps by Shakespeare's 'weird (wyrd) sisters', the three witches featured in Macbeth who prophesy the protagonist's forthcoming doom. Whilst we now use the word weird for all sorts of everyday occurrences, it is this more supernatural bent that is in keeping with the use of the term in this book.

Most mainstream scientific journals and publications assume that weird phenomena are not real, and, because of their adherence to the brain-based view of consciousness, *cannot* be real. For the brain-based view to be meaningfully challenged assumes, of course, that there are no other adequate explanations for these experiences. Those who support a brain-based view of consciousness maintain that all apparently weird phenomena can be accounted for by some form of normal explanation meaning that, on closer inspection, they are not ultimately weird. This would mean that these phenomena—despite appearances—remain consistent after all with the idea that the brain produces consciousness. We'll examine these sceptical perspectives as we go along.

The topic of this chapter is weird phenomena that seem to suggest that consciousness may be more than just our brains. Examples include instances where people express themselves lucidly despite having a severe brain impairment that would seem to make that impossible, and savant syndrome where a person suddenly and inexplicably acquires a new skill such as being able to understand and speak a foreign language, often because of some trauma such as a blow to the head. There are also phenomena that seem to point to consciousness surviving the death of the physical body such as those who are bereaved seemingly having contact with deceased loved ones as well as age-old reports of ghosts and apparitions. In this chapter, we'll focus on the evidence for just two kinds of such weird phenomena—where someone's consciousness is seemingly able to explore worlds beyond their body and reincarnation, where a person believes that they have previously lived a life as someone else before they were born.

Travelling Away from the Body

Throughout history, people have reported that their consciousness has seemingly travelled outside of their body so that they can perceive places, events and objects that should be beyond the reach of their senses. Such experiences are known as astral travel or out-of-body experiences (OBEs). To get a sense of what these are like, it is worth quoting at some length how the author Gene Hart describes the first time he felt his consciousness leaving his body:

> One night, I found myself waking up... I stayed still and immediately began spontaneously feeling trembling vibrations all over my body, like an earthquake; I started to see colours swirling in my vision through my closed eyelids. I heard noises, like a mix of a hurricane and white noise...
>
> I floated upwards. It was a slow and intense separation. As I gradually detached, I saw my physical body fast asleep beneath me. This new type of ghost-like body, which I had never experienced before, felt acutely energetic... I felt blisteringly alive, but at the same time, heavy and more real than anything I had ever experienced before in my waking life... I felt the weight of my energetic body lower to my bedroom floor, and I felt the tactual sensation of the carpet under my feet. I looked around my dark lit room, which felt more natural and vivid than I'd ever seen before; it had a luminous and magical quality to it. I was awestruck... I was indeed separate from my physical body and brain as I observed it sleeping in my bed; there was truly no doubt of this...
>
> With the knowledge that this dimension was ultimately non-physical, I flew through my bedroom wall and came out hovering on the other side to see a beautiful clear night sky with a full moon shining its light on the lush treetops. I had never seen outside my mundane house with such a spiritual and magical quality before. It was visually stunning, but what's more was the sense of vibrancy, life and emotion that emitted from everything I saw. I became so overwhelmed with what I was seeing that I woke up. I was in a state of shock and didn't go back to sleep that night. However, I was also overjoyed. I truly felt as though I had found some hidden treasure, stored on the edge of reality, located parallel to the physical dimension.[59]

It is tempting to write off such experiences as elaborate dreams or (literally) flights of fancy. One person to do so is Chris French, psychologist and prominent sceptic. He argues that OBEs are not a perception of consciousness leaving the body but instead are simply products of the imagination and/or a "disruption of the neural processes responsible for integrating information from a range of sensory modalities".[60] In other words, these occur when the brain misinterprets what it's receiving from the senses.

To support this contention, French cites studies where people can be made to believe that their consciousness is somehow separated from their body. One well-known experiment he describes involves hiding a volunteer's hand from view and replacing it with a fake rubber hand on the table in front of them. The experimenter gently strokes both the obscured real hand and the fake hand with a soft brush. After a short period of time, the volunteer usually reports that the fake hand feels like their real hand to the extent that when the experimenter goes to strike it with a hammer, they react exactly as if their real hand is in danger. Whilst French concedes that this experience is far from replicating anything approaching most reports of OBEs, he describes other studies that can make volunteers feel like their body is several feet away. For example, researchers Valeria Petkova and Henrik Ehrsson used virtual reality to encourage participants to experience a sense of 'swapping bodies' with a dummy to the extent that "not only... their sense of self was located inside a mannequin but even that they were shaking hands with themselves!".[61] However, the issue with these experiments is that they do not seem to resemble actual reported OBEs such as Gene Hart's vivid experience related above. As my colleague Chris Roe puts it, the "description of the supposed OBE in these experiments is frustratingly impoverished and offers little confidence that this could account for some of the rich and complex OBEs that have been reported in the real world".[62]

Reports of OBEs that are even more challenging to explain as brain-based hallucinations are those that involve someone seeing or hearing something when away from their body about which they should have no normal way of knowing and which is later confirmed as accurate. Robert Monroe, a radio broadcasting executive who in the 1970s wrote several books about his extensive experiences with OBEs, describes instances of this type in *Journeys Out of the Body*.[63] For example, one morning, Monroe had an OBE in which:

> The first thing I saw was a boy walking along and tossing a baseball in the air and catching it. A quick shift, and I saw a man trying to put something into the back seat of a car, a large sedan. The thing was an awkward-looking device that I interpreted to be a small car with wheels and electric motor. The man twisted and turned the device and finally got it into the back seat of the car and slammed the door. Another quick shifty and I was standing beside a table. There were

> people sitting around the table, and dishes covered it. One person was dealing what looked like large white playing cards around to the others at the table. I thought it strange to play cards at a table so covered with dishes, and wondered about the overlarge size and whiteness of the cards.

Monroe tells of how that evening he visited his friends, Mr and Mrs Bahnson, who were aware of his forays out of the body. He writes that

> on a sudden hunch, I knew the morning event had to do with them. I asked about their son, and they called him into the room and asked him what he was doing between eight-thirty and nine that morning. He said he was going to school. When asked more specifically what he was doing as he went, he said he was tossing his baseball in the air and catching it. (Although I knew him well, I had no knowledge that the boy was interested in baseball although this could be assumed.) Next, I decided to speak about the loading of the car. Mr. Bahnson was astounded. Exactly at that time, he told me, he was loading a Van DeGraff generator into the back seat of his car. The generator was a large, awkward device with wheels, an electric motor, and a platform. He showed me the device. (It was eerie to see physically something you had observed only from the Second Body.) Next, I told about the table and the large white cards. His wife was excited at this one. It seems that for the first time in two years, because they had all arisen late, she had brought the morning mail to the breakfast table and had passed out the letters to them as she sorted the mail. Large white playing cards! They were very excited over the event, and I am sure they were not humoring me.[64]

Whilst there are a plethora of videos and books offering techniques to induce OBEs with titles such as *Navigating the Out-of-Body Experience*[65] and organisations purporting to train OBEs such as the institute established by Robert Monroe,[66] many occur involuntarily when people are very ill and close to death. One oft-cited instance is that of Maria, a migrant worker in the United States who was rushed to hospital with a heart attack whilst visiting friends in Seattle. When she was on the coronary care unit, she experienced a cardiac arrest. After she had been resuscitated, she told her critical care social worker Kimberley Clark that during the episode her consciousness had left her body. Maria said that at one point she had found herself floating high up outside of the hospital and noticed a tennis shoe on the ledge of the third floor. She was able to describe exactly where she saw it along with some details about its appearance, including one of its laces being stuck to the sole and the area around the little toe being worn. Despite being sceptical that what Maria had told her could really mean that her consciousness had travelled out of her body, Clark investigated and found a tennis shoe just where Maria had said she had seen it. Perhaps more remarkable still was that it wasn't possible to ascertain from the hospital window whether it had the features Maria had described and yet, when the shoe was retrieved, the lace and the worn area corresponded exactly to her account.[67]

Another example of what is known as a 'veridical OBE' (meaning that what is reported can later be verified) involved a severely-ill patient who said that whilst out of their body they were seemingly able to see a section of their hospital ward that wasn't visible from where they were stationed:

> I was in the hospital having had an operation for peritonitis; I developed pneumonia and was very ill. The ward was L-shaped so that anyone in bed at one part of the ward could not see around the corner. One morning I felt myself floating upwards, and found I was looking down on the rest of the patients. I could see myself propped up against pillows, very white and ill. I saw the Sister and nurse rush to my bed with oxygen. Then everything went blank.
>
> The next I remember was opening my eyes to see the Sister bending over me. I told her what had happened but at first she thought I was rambling. Then I said, 'There is a big woman sitting up in bed with her head wrapped in bandages and she is knitting something with blue wool. She has a very red face.' This certainly shook her as apparently the lady concerned had a mastoid operation and was just as I described. She was not allowed out of bed and of course I hadn't been up at all. After [providing] several other details such as the time by the clock on the wall (which had broken down) [and was also not visible from the patient's bed] I convinced her that at least something strange had happened to me.[68]

There have been many other such incidents recorded. The researcher Madelaine Lawrence, for example, reports two cases involving patients who had OBEs whilst unconscious and could accurately describe items of clothing worn by medical staff during this period, in one case distinctive plaid shoelaces worn by an intensive care unit nurse and in another a bright yellow smock worn by a respiratory therapist.[69] These reports provide evidence that consciousness is apparently functioning outside of the body. The very possibility of being able to sustain a vivid, complex and coherent experience at a time when brain functioning is severely compromised such as during general anaesthesia or cardiac arrest poses a formidable challenge to brain-based theories of consciousness.

Living a Previous Life

Another type of weird phenomenon suggesting that consciousness might be able to exist beyond the brain is reincarnation. Reincarnation refers to a non-material essence of a person returning to another physical body after death. A number of religious and spiritual traditions subscribe to the belief that life is an ongoing cycle of birth and rebirth. This is particularly widespread in Indian philosophies such as Hinduism and Buddhism where it is thought that an individual's soul (*atman*) is continually reborn until they realise the essential oneness of absolute reality (*brahman*) and so achieve an escape from this cycle.

Some claims of people apparently remembering a previous life or lives have been made as a result of hypnotic regression. This technique involves a therapist hypnotising a participant and then asking them to recall events at progressively earlier stages in their current life until past memories reach back beyond their birth and seem to come from a supposed previous life. One hypnotherapist who regularly practised this type of regression was Arnall Bloxham. Out of more than four hundred cases, perhaps his most notable was that of Jane Evans (not her real name), a Welsh housewife who was featured in a 1975 BBC documentary on the subject. Jane claimed to recall memories from six previous lives including Sister Grace, a nineteenth-century Catholic nun living in Iowa; Anna, a sixteenth-century lady-in-waiting to Catherine of Aragon who was brought to England to marry Prince Arthur; and Livonia, who lived in third-century Roman York. According to experts consulted for the documentary, Jane was able to recall details about these lives that bore a remarkable resemblance to historical facts, although others suggested that they were also similar to the contents of novels that Jane might have read.[70]

One significant issue with this approach to investigating reincarnation is that the use of hypnotic regression has been heavily criticised for not producing accurate recall, for example when it's been used for eyewitness testimony in criminal cases or where therapists have recovered supposedly repressed memories of childhood abuse later found not to have occurred.[71] It seems to be all too easy for therapists inadvertently to implant false memories or for participants to recall details from books or films rather than actual memories. Research into past life regression specifically has found that this can indeed take place. For example, the psychologist Edwin Zolik regressed a man who supposedly remembered details of his previous life in the Irish

Guards who died when thrown from his horse, a story, it emerged from further investigation, that he had originally heard from his parents.[72]

A different approach to investigate whether reincarnation is a real phenomenon is to examine the claims of people who seem unprompted to have memories of having lived a previous life. Whilst such memories can happen at any age, it is most commonly associated with young children when they are old enough to begin to be able to express themselves. Typically, the child in question may declare that they do not belong in this life, that they have a different name and that they have family members including sometimes their own spouse and children in addition to or instead of their current immediate relations. In the late 1950s, the researcher Ian Stevenson began systematically investigating such cases. He and his associates amassed a database of over 2,500 of these cases of apparent reincarnation.[73]

One case Stevenson documented relates to twin girls born in 1958 in the north east of England who were named Gillian and Jennifer Pollock.[74] Their parents claimed that the girls were reincarnations of two other of their children, Jacqueline and Joanna, who had died tragically in 1957 when they were hit by a car driven by a woman taking her own life. Among the apparent indications of reincarnation noted by the parents were:

- The twins had a phobia of cars.
- Gillian was once seen cradling Jennifer's head, saying, "The blood's coming out of your eyes. That's where the car hit you". The father recalled that when he identified the bodies, Jacqueline's head had been bandaged above the eyes.
- Gillian once pointed to a birthmark on Jennifer's forehead and said, "That is the mark Jennifer got when she fell on a bucket", which had indeed been the cause of a scar Jacqueline had had on her forehead.
- Jennifer had another birthmark on her left waist, which was in the exact same place as where Jacqueline's birthmark had been.
- The twins seemed to be familiar with some areas of the town of Hexham; for example, knowing their way to the park, despite the family having left the area when they were very young, suggesting that they had perhaps inherited Jacqueline and Joanna's memories.

- Gillian and Jennifer recognised toys that had belonged to Jacqueline and Joanna and said correctly that they'd been Christmas presents.

Chris French notes that there are a number of issues with such accounts. Firstly, there is an understandable tendency on the part of researchers only to report the strongest cases. The many instances where cases are weaker or where children don't describe any past life are naturally not included so there is an inherent selection bias in the reincarnation literature. Secondly, the details of the cases that are reported rely on the people involved accurately recalling what was said by the children or events that had happened to them. However, such testimony is notoriously unreliable, for example because people frequently misremember, are prone to exaggeration and those involved also tend only to report features that seem to support reincarnation and not the features that don't. This is particularly so when there is a desire on the part of the parents to believe that their children have had previous lives. It is especially telling that in the case of the Pollock twins their father John had a strong belief in reincarnation and had prayed for proof of this phenomenon.

Many of the cases that are reported by Stevenson and other researchers involve members of the child's family claiming to have found a person who died before the child was born who closely matches what the child has been saying about them. A major stumbling block for these cases is that such evidence has already been identified before researchers have begun investigating meaning that it is difficult to verify conclusively whether these claims are genuine. It's not hard to imagine the possibility that what the children were said to have reported about their previous lives might have been subsequently embellished or invented—entirely or in part—in order to fit the facts on the ground and so make the case appear more impressive. This may not necessarily be done deliberately but could simply be a result of wishful thinking or those involved looking for evidence to support their pre-existing beliefs, which is known as confirmation bias. Therefore, in terms of providing stronger evidence for reincarnation, it is preferable for as many details as possible of what the child is reporting to be thoroughly documented before the case is fully researched and the facts can either be confirmed or rejected.

Michael Nahm calls such cases 'before-cases' and states that he knows of thirty-one that have been published in the literature.[75] One of the strongest of these is that of Ryan Hammons, a boy born in

Oklahoma in the United States in 2004, a case investigated by Jim Tucker, Stevenson's successor.[76] Whilst Ryan was unable to speak until he was four because he had enlarged adenoids, once he had had these removed, Ryan began telling his parents (Cyndi, who was the deputy county clerk, and Kevin, a police lieutenant) of having lived a life before he was born. He told them that he was from Hollywood and asked to be taken there. He recounted many memories that seemed to suggest he was someone else and, acting on a suggestion from Kevin, Cyndi began taking notes of what Ryan was saying. Cyndi also borrowed books about Hollywood from their local library to show to Ryan. One of the books included a still from the 1932 film *Night After Night* in which Ryan pointed to an actor who he said was his past self. Unfortunately, the caption for the picture did not include the man's name and Cyndi and Kevin were unable to identify who he was from the movie credits.

In 2010 when Ryan was six, Cyndi wrote to Jim Tucker having heard of his research to see if he could help. As chance would have it, one month prior to receiving Cyndi's letter, Tucker had been contacted by a cable television company wanting to make a documentary about reincarnation for the series *The UneXplained*. This case provided an opportunity to document on film what Ryan was saying before conducting further research that might confirm or discount his statements. The company recorded interviews with Ryan and also hired an archivist who, after a fair amount of investigative research, was able to establish that the actor identified by Ryan might be a person called Marty Martyn who seemed to be a good fit for some of Ryan's descriptions.

Some of the many statements about Marty that Ryan made that were documented and later found to be correct included: he was very rich and lived in a big house; he had three adopted boys and one daughter for whom he bought a dog when she was six, which she didn't like; his house had a swimming pool; his mother had curly brown hair; he drove a green car whilst his wife's car was black; he ran a talent agency; he knew Rita Hayworth and Marilyn Monroe; and he died at 61. Ryan was also able to state within a year when Marty was born saying, "I'm not five; I'm closer to 105 when I was here before". When he made this statement, it had been 106 years since Marty's birth. By 2016, Cyndi had recorded 230 individual pieces of information provided by Ryan. Fifty-five (24%) of these could be confirmed, 15 (6.5%) were found to be incorrect or at least unlikely and 140 (70%) were unverifiable.

Ryan's personality and behaviour also seemed to bear similarities with what was known about Marty. For example, he asked to dress in what he called 'agent clothes' such as suits, dress shirts and ties. At his fourth birthday party he initiated a game where the other children would act in a movie he was directing. Like Marty, he had a fear of cats. He said that he liked Chinese food and, when he was first taken to a Chinese restaurant, he was able to use chopsticks spontaneously despite not being shown how to do this, much to his parents' surprise. The reincarnation aspect of his life ended in 2011 when Ryan was seven, soon after the documentary was televised, when he told Cyndi that it was time for him to stop talking about Marty and begin a more regular childhood. However, elements of Marty's personality seemed to persist in Ryan such as his liking for 1950s music, being supportive of Republican politics and expressing an interest in Judaism, despite this conflicting with his parents' Christian faith.

***

The weird phenomena of out-of-body experiences and reincarnation seem to point to consciousness being able to exist beyond the brain. It is difficult to reconcile such weird experiences—if they are as they appear to be—with the view that consciousness is entirely produced by the brain. This is why most theorists of a brain-based consciousness do not take such accounts seriously, if they acknowledge them at all. In the next chapter, we'll examine another category of phenomena that seems to cast doubt on the brain-based paradigm and points to consciousness being able to transcend apparent physical constraints of space and time.

Chapter 4

Can Consciousness Transcend Space and Time?

If consciousness was entirely brain-based, as the prevailing scientific view would have it, information about the world outside of us would only be available through our senses of seeing, hearing, touch, taste and smell. However, people report that they sometimes obtain information directly from others or from the future without recourse to any of the known senses or inferring from what is already known. These phenomena are known as extrasensory perception (ESP) and they seem to demonstrate that consciousness is able to transcend space and time. Such experiences are challenging, perhaps even impossible, for the brain-based theories we discussed in Chapter 2 to account for. But what is the evidence that ESP is real and not just a figment of the imagination, wishful thinking or just plain made up? In this chapter, we'll look at the evidence for mind-to-mind communication and knowing the future.

Connecting Minds

Telepathy refers to direct communication between minds and its literal meaning is 'feeling at a distance'. One story of telepathy told to me personally was of a friend of mine who had gone hiking alone in the winter in mountains close to where she was living, leaving her partner at home with the children.[i] Although she was experienced

---

i Some details of this anecdote have been changed to protect anonymity. However, the essence of the story is accurate.

with these kinds of conditions and regularly went on such outings, she slipped on some ice and hit her head on a rock, rendering her unconscious. Luckily she was on a well-trodden footpath and it wasn't long before some other walkers found her and called the local mountain rescue. She was airlifted by helicopter to the nearest hospital. What makes this remarkable is that her partner experienced a strong sense of foreboding some time before she was contacted by anyone about the accident. The feeling was so intense that it caused her to pack a bag and make arrangements with a neighbour to take care of their children, enabling her to be with her partner much earlier than would otherwise have been the case. Somehow, she just knew that her partner had experienced a serious trauma. Fortunately, my friend was able to make a full recovery and live to tell the tale.

Accounts of telepathy are relatively widespread. One recent survey found that around 10% of the UK population—some five million people—said that they had had such an experience.[77] Most instances are much less dramatic than what happened to my friend. For example, I sometimes notice thinking of something completely unrelated to anything going on around me only to have my partner express the very same thought (interestingly, this rarely happens the other way around). Just as with cases of out-of-body experiences and reincarnation, which we looked at in the last chapter, reports of telepathy are also potentially subject to the vagaries of memory, exaggeration or even people making stuff up to add spice to their anecdotes. So how might we determine whether telepathy is a real phenomenon? The answer comes from conducting scientific experiments that are designed to rule out other possible explanations.

Although there have been telepathy tests performed at least since Victorian times, probably the first person to construct large-scale and formal experiments on ESP was Joseph Banks Rhine in his work at Duke University in the United States between the 1930s and 1960s. Initially trained as a botanist, he and his wife Louisa Rhine (who also pioneered research in this area) had become interested in what was then usually called psychical research, the science of investigating weird phenomena, now more commonly known as parapsychology. They conducted investigations of mediums in the Boston area who claimed to be in contact with the deceased and exposed at least one who they found used fraudulent means. This work convinced Rhine that it was important to include rigorous controls and he wrote to one colleague saying, "I have decided not to accept any supernormal psychical phenomena without actually obtaining them under my own conditions, and witnessing them myself".[78] He went on to devote the rest of his career to the scientific investigation of various weird (paranormal) phenomena.

Most famously, Rhine worked with the psychologist Karl Zener to develop cards with five distinct and easily remembered images to be used as targets for extrasensory information transfer. These are known as Zener or ESP cards and have become quite recognisable through being featured in numerous television programmes and films, for example in the original *Ghostbusters* movie.[79] Each card shows a simple figure of either a circle, a cross, wavy lines, a square or a star. The basic design of Rhine's telepathy experiments was for one person to act as a sender who attempts to transmit mentally one of these five images to someone else acting as a receiver. These participants are usually situated at some distance from each other and so should be unable to know what is being transmitted to them by any means other than telepathic (mind-to-mind) communication. A typical run in Rhine's experiments consisted of twenty-five trials. If telepathy is not real, we should expect the results from a large number of trials to be what we'd get from just guessing, that is about five symbols correct in each run, or what is known as a hit rate of 20%.

In one early set of experiments involving two psychology graduate assistants as sender and receiver, 1,950 trials were carried out.[80] There was an electric fan whirring that was intended to obscure any intentional or unintentional signalling that might occur between the participants. Whilst getting five images correct out of 25 (20%) is what we'd expect by chance, this pair scored an average of 14 hits out of 25 (56%) when they were in the same room, 14.6 (58%) when next door to each other and 16 (64%) when two rooms away. In one of the runs, there was a perfect score of 25. The likelihood of these results happening simply by chance is astronomically small—you'd expect this result to occur just once in 4×10^{295} similar experiments (the number 4 with 295 zeroes).

Most of Rhine's experiments achieved results that weren't quite as impressive, although his cumulative findings were still highly unlikely to have occurred by chance, supporting an ESP hypothesis. Results such as these led a number of scientists at the time to conclude that telepathy was real. One was Alan Turing, famous for making vital contributions in the second world war by deciphering encrypted messages sent by the German military and who many credit with the birth of modern computer science. In 1950 Turing published what is regarded as a seminal paper in the development of artificial intelligence, describing what he calls 'the imitation game'.[81] This is a test to determine whether a computer is sufficiently advanced to fool someone who converses with it into believing that it is human rather than a machine. One possible issue with the test, Turing suggested, was that it might be possible for telepathy—the capacity for which he suggested was a key difference between humans and machines—to get in the way of accurate results from this test. He wrote of ESP that "these disturbing phenomena seem to deny all our usual scientific ideas. How we should like to discredit them! Unfortunately the statistical evidence, at least for telepathy, is overwhelming".[82]

Whilst the methods Rhine used were more rigorous than most of the experiments that had come before, critics have highlighted a number of flaws meaning it could conceivably have been possible for such results to be achieved by means other than telepathy. For example, Robert Carroll has noted that card shuffling is not sufficient for true randomisation and that the odds of making a correct guess change each time a card is removed from the pack.[83] Terence Hines suggests that in many of Rhine's experiments, the shielding of the cards wasn't always sufficient to rule out non-telepathic means of knowing because the receiver might have been able to see an image of the card reflected in the sender's glasses or even in their eyes.[84] Another sceptic, Jonathan Smith, points out that there might have been indentations on the backs of the cards from the printing process meaning that the receiver might be able to see what was on front of the card, even if only subliminally.[85]

As well as suggesting that many accounts of supposed telepathy might be due to misremembering or exaggeration, Chris French proposes that they could also be explained by coincidence in what he calls the 'law of truly large numbers'.[86] He suggests that, in many cases, striking correspondences between two or more events that we might be tempted to interpret as paranormal are in fact explainable by the often countless opportunities there are for such coincidences to occur. This means that such weirdnesses are bound to happen every now and again. One example French highlights is when we find ourselves thinking about someone we've not seen for ages and then receive a telephone call from them at that very moment. Rather than telepathy, French suggests this can be explained in the following way:

> Just consider how many people you think about in a single day. Most of the time when you think about them, either the phone does not ring or the phone rings and it is someone else calling. But, occasionally,

simply based on the laws of probability, the person calling will be the person you were thinking about.[87]

Parapsychological experiments are designed to rule out such non-paranormal explanations. In particular, the specific phenomenon of 'telephone telepathy' described by French has been the focus of experiments originally conceived by Rupert Sheldrake, whose theory of consciousness we'll examine in Chapter 9. Sheldrake's first studies of telephone telepathy involved five people for each phone call.[88] One of this group is designated to receive phone calls and, for each trial, one of the other four is randomly selected to make the call. As the phone is ringing the person receiving the call states who they believe is at the other end of the line (without the phone displaying caller ID, of course). If nothing telepathic is going on and the receiver is relying entirely on guesswork, then the chances of correctly identifying the caller is one in four, or 25%. In fact, over the course of these experiments involving almost 600 phone calls, the receiver correctly identified the caller an average 40% of the time. The odds against this result happening by chance are around one in one quadrillion (one with fifteen zeroes). Not everyone taking part scored above chance, meaning that those who did so were correct at a higher rate even than this. An analysis of 23 similar studies conducted up to 2024 by a number of different researchers showed a combined success rate that on average was nearly 9% greater than expected from chance, strongly supporting the idea that something weird is going on.[89]

Although most telepathy experiments have been conducted using human participants, many people have reported a telepathic connection with animals to whom they are emotionally bonded such as their pets. This has been of particular interest to Sheldrake who conducted experiments involving a mongrel terrier called Jaytee, who seemed to know telepathically when his owner Pam Smart was coming home, a story that was widely covered in the media at the time.[90] Jaytee had been observed by Pam's parents to go to the window and stay there when Pam was coming home even though they weren't otherwise aware that she was on her way.

In order to test whether Jaytee was acting telepathically, Sheldrake's experiment involved videotaping Jaytee whilst Pam was away and when she was returning. Each recording was analysed to see how much time Jaytee spent by the window during three separate periods—when Pam was away, the ten minutes leading up to when she was about to return and the actual period when Pam was on her way home. If nothing telepathic was occurring we should expect to see no real pattern in the data. However, in the twelve experiments when Pam returned at a randomly-determined time, Jaytee was at the window 4% of the time during Pam's absence, 23% in the pre-return period and 55% when she was on her way. In another thirty experiments, Pam returned at a time of her own choosing but kept this from her parents who were at home looking after Jaytee. This yielded similar results with Jaytee being at the window 11% of the time whilst Pam was away, 31% of the time when Pam was about to come home and 65% during the actual return period. Statistical analysis shows that the chances of Jaytee just happening to spend these varying amounts of time at the window by chance is less than one in 10,000.[91]

Sheldrake maintains that the design of this experiment rules out several possible non-weird explanations that are often put forward by sceptics about this phenomenon. The results can't be due to Jaytee knowing Pam's routine because the experiment involved her returning at different times of the day and evening. Jaytee can't be responding to cues from those at home because Pam's parents weren't aware of her return time. They can't be due to selective memory because all of the experiment sessions were recorded and analysed. Jaytee didn't just visit the window more often as time went by because the data show that this wasn't the case. Finally, Jaytee can't be responding to hearing Pam's car because Jaytee went to the window before Pam had even set off, he behaved the same way even when she returned in unfamiliar vehicles such as a taxi, and her car was between 7 and 25 kilometres away before she returned making it implausible that Jaytee could have heard the car over other noises closer by. Sheldrake has commented that the advantage of involving dogs rather than humans in such experiments is that they remain enthusiastic and don't get bored no matter how many times an experiment is conducted.

Seeing the Future

Whilst telepathy suggests that consciousness can transcend space, some people have also reported instances where consciousness appears to defy our usual notions of time in the form of visions of the future that later turn out to be uncannily accurate. Here's one instance of what is known as precognition from the autobiography of the famous author Mark Twain (1835–1910) regarding his brother Henry:[92]

> In the morning, when I awoke I had been dreaming, and the dream was so vivid, so like reality, that it deceived me, and I thought it *was* real. In the dream I had seen Henry a corpse. He lay in a metallic burial case. He was dressed in a suit of my clothing, and on his breast lay a great bouquet of flowers, mainly white roses, with a red rose in the center.

When Twain awoke, understandably shaken, he went with foreboding to the house he'd seen in his dream and found to his relief that Henry had not died. He wrote:

> There was nothing real about this... I can still feel something of the grateful upheaval of joy of that moment, and I can also still feel the remnant of doubt, the suspicion that maybe it *was* real, after all.

Just a few weeks later, Henry was tragically killed after being badly burned in a boiler explosion on the Mississippi boat where he was working. Twain went to the house of his dream and which he had visited previously.

> The coffins provided for the dead were [usually] of unpainted white pine, but in this instance some of the ladies of Memphis had made up a fund of sixty dollars and bought a metallic case, and when I came back and entered the dead-room Henry lay in that open case, and he was dressed in a suit of my clothing. He had borrowed it without my knowledge during our last sojourn in St. Louis; and I recognized instantly that my dream of several weeks before was here exactly reproduced, so far as these details went—and I think I missed one detail; but that one was immediately supplied, for just then an elderly lady entered the place with a large bouquet consisting mainly of white roses, and in the centre of it was a red rose, and she laid it on his breast.

A more recent instance of precognition that seems to have served as a warning from the future is recounted in the book *The Premonition Code* by Theresa Cheung and Julia Mossbridge.[93] This is from a woman named Rachel, who, like Twain, told of experiencing a vivid and upsetting dream.

> I was playing with my daughter in a house made of glass. We were playing catch with a ball and laughing. I started to hear birds singing so loudly I got distracted and I missed catching the ball my daughter threw to me. The ball flew into the glass and the house shattered with splinters and shards everywhere. I heard my daughter screaming. I woke up absolutely terrified and covered in sweat.

A few weeks later, Rachel's daughter was invited to a friend's house for a party. The children were playing in the conservatory. She wrote, "it was like the glass house of my dream. I heard birds singing loudly. I thought of my dream and told the girls to come with me into the kitchen". A few moments later a cricket ball flew from the garden next door smashing into the glass, which shattered everywhere. She concludes, "I shudder to think what might have happened if the girls had still been playing there".

Many people report similar experiences. The same representative survey quoted above in relation to telepathy found that 13% of the UK—around seven million people—said that they have had instances of precognition.[94] Again, we need to ask whether people are just misremembering their dreams and visions, embellishing their stories after the fact, whether consciously or unconsciously or whether Chris French's 'law of truly large numbers' is at play. It is to rule out such non-weird explanations that we need to perform scientific experiments to assess whether consciousness can indeed transcend time. Parapsychologists have devised a number of different ways in which we might test this hypothesis.

One approach is to detect whether unconscious processes in the body respond to future events before they occur. (It's not hard to envisage that having some unconscious foreknowledge of a future event would be very helpful for an organism's survival meaning that precognition might be an evolutionary adaptation.) This type of experiment focuses on what is known as presentiment—the unconscious pre-sensing of a stimulus through some physiological expression. Presentiment experiments have been conducted that measure various autonomic functions over which most people have no conscious control such as skin conductance, heart rate or the amount pupils dilate. One useful aspect of this type of experiment is that measurements are made by instruments so making results less open to subjective judgement.

In these experiments an image is randomly selected by a computer that is one of two distinct types that are each known to produce different responses in people, such as an image of something calming, like a landscape, or something that evokes an emotional response, like a photo of a heated argument. Measurements are taken throughout the experiment, including the period leading up to when the computer randomly selects an image, to see whether the body is making some unconscious response to this future event. Results show that people's physiology does indeed seem to respond ahead of time differently to different types of images, even when measurements are taken some seconds prior to when the computer makes the selection.

For example, the pupil might dilate more before being shown a picture of someone visibly upset than a restful sunset, even though the choice of that image by the computer has yet to be made. Leading researchers in this area, Julia Mossbridge, Patrizio Tressoldi and Jessica Utts, have conducted a review of 26 such studies that took place between 1978 and 2010 and found combined results that would only be seen by chance once in 3×10^{12} times.[95]

Other scientists have focused on whether people might be able to make choices based on something that has yet to occur. In 2011, a paper was published in a leading psychology journal by Daryl Bem that detailed nine precognition experiments.[96] These experiments involved various set-ups where participants were asked to press one of two computer keys in response to an image or other information that was about to be randomly selected and shown on the screen. All of the experiments gave positive results showing that people seemed to be able to sense future events that had not yet happened. The interesting thing about Bem's experiments is that they were based on standard procedures that have been in use for many years to test established psychological effects, such as priming. Bem just flipped the sequence of stages in the experiment so that participants' choices were made before images were selected in order to test precognition.

The paper created a storm in the world of psychology and beyond. Chris French, working with Stuart Ritchie and Richard Wiseman, attempted to replicate one of Bem's experiments but, unlike Bem, they found no evidence of precognition. When they tried to publish their results in the journal that had published Bem's study, the editor declined, saying that their policy was not to publish replications. Submissions to other journals failed for similar reasons, although they did eventually manage to publish the article.[97] This led some to wonder whether there was a problem with how science was being conducted. In 2005, John Ioannidis had published an article in a leading medical journal titled "Why Most Published Research Findings are False", pointing out that issues such as intense competition to secure funding and scientists' in-built biases meant that a lot of mainstream research carried out in our universities and institutions might be leading to mistaken conclusions.[98] Bem's high-profile research seemingly showing that precognition was a real thing supercharged this issue. One article in an online magazine had the headline "Daryl Bem Proved ESP is Real. Which Means Science is Broken".[99]

In order to assess the extent of the problem in fields other than parapsychology, Brian Nosek and colleagues set about trying to replicate 100 published psychology studies. Shockingly, only 36% of these were found still to produce positive results and then at smaller effect sizes.[100] There seemed to be a problem with medical research too. A team of over 200 set out to repeat 193 published cancer research experiments, of which they were only able to replicate with similar results around a quarter, 85% of which had smaller effect sizes than those originally found.[101]

As a result of what has become known as the 'replication crisis',[102] there is now a focus on approaches such as 'registered reports' where details of an experiment including its methodology and statistical techniques are written up and (if good enough) accepted for publication by a journal before the experiment takes place. This has two advantages. Firstly, this should lead to a more balanced picture of scientific findings in the literature because it commits journals to publishing studies regardless of whether results support hypotheses or not. Some journals have in the past pursued a policy of only publishing novel papers, meaning that negative findings and replications often failed to see the light of day, as was the case with French and colleagues' attempted replication of Bem's experiment. Secondly, registered reports can help prevent what are known as 'questionable research practices' (QRPs). QRPs are usually inadvertent and rarely involve actual fraud, but they nevertheless mean that the quality of research can be suboptimal and lead to findings that are not reliable. For example, researchers have sometimes been known to change the way they present their findings after the fact to make them appear more favourable, to create hypotheses in retrospect to fit the data and to try out many different analyses on their data to uncover positive findings in amongst largely negative results. Registered reports, along with preregistration, where researchers publish their methods in an online repository, commit researchers to applying the methods and analyses they said they would use prior to carrying out the research.

It has become accepted wisdom that the idea for registered reports dates back to 2012 when it was suggested independently by two academic journals in the fields of medicine and psychology.[103] However, this approach was actually first proposed in 1975 by the parapsychologist Martin Johnson and soon after adopted by the *European Journal of Parapsychology*.[104] In fact, parapsychology has often been ahead of the curve in devising modifications to address concerns raised by critics that have gone on to find application in other disciplines. For example, randomisation has been in use in medical research since the 1930s to allocate patients into treatment and control groups but this technique was first pioneered in telepathy experiments conducted in the late nineteenth century.[105]

Bem's precognition research has had successful replications as well as unsuccessful. A 2016 review of 90 experiments conducted by researchers from 14 different countries found that their collective results supported his positive findings.[106] However, the issue with such reviews that combine many similar studies, which are known as meta-analyses, is that any QRPs that may be present in individual studies can be obscured and even compounded when lumped together. A more reliable and robust alternative to meta-analysis is to conduct a single large experiment with exacting standards. This approach was used to attempt to replicate one of Bem's experiments. This project involved a large team of researchers made up of those who were both open to the possibility of precognition and those who were more sceptical. The experiment failed to find evidence that participants could 'feel the future'. The paper concluded that "even if ESP [does] exist, our findings strongly indicate that this particular paradigm... is unlikely to yield evidence for its existence".[107] On reflection, this perhaps should not be surprising. After all, as we saw with Mark Twain's and Rachel's experiences of seeing the future, precognition is often reported as occurring within dreams and only rarely when someone is wide awake and involved in a mundane task such as operating a computer as participants in Bem's study were asked to do.

The idea of trying to emulate the conditions under which ESP happens in the real world, such as when people are in a dream-like state, is the motivation behind the use of the ganzfeld. A ganzfeld experiment is designed to help participants enter a sleepy state through their lying comfortably, immersing their vision in diffused red light and playing them a relaxing neutral sound, usually white noise. The intention is to reduce the conscious mental chatter that might get in the way of experiencing ESP. A typical set-up to test precognition involves the participant reporting whatever impressions come to mind that might be related to a future target, typically an image or a video clip. After the session is complete and the recording of what the participant said is 'locked in' so it can't be changed, the target image or video clip is randomly selected and shown to the participant.

If precognition is really happening, at least some of what the participant reports should be related to the target image or video. At the end of the session, the participant looks at a small number of possible targets, including of course the correct one, and decides which is the one that was the randomly selected target. A variation is that someone independent, who doesn't know the correct target, will look at a transcription of what the participant said and make this decision instead. Either way, we compare how many correct hits we get with what we would expect with chance. For example, if there are four possible targets then the likelihood of a hit is 25%. Patrizio Tressoldi and Lance Storm reviewed 78 experiments that used this type of design that were conducted between 1974 and 2020. This meta-analysis involved almost 3,000 separate trials and found an average hit rate of 33% compared with the chance rate of 25%.[108] The likelihood that this could be down to chance is just one in 5×10^{22}.

However, as we've seen, the use of meta-analysis has been called into question. With this in mind, researchers Caroline Watt and Abby Pooley led a large ganzfeld study to test precognition. The study was preregistered, which means every aspect of the experiment was publicly declared before it took place so as to rule out possible QRPs. There were 240 participants, which is a lot considering that each trial takes around two hours to complete. They achieved a hit rate of 30%. Watt and Pooley took great pains to ensure there were no artefacts in their design meaning that something other than precognition could have caused this higher-than-chance hit rate. For example, they ran the software they used over 5,000 times in an empty room without a participant. The hit rate was exactly 25%. They checked the random number generator to ensure it was truly random. They checked the records of each trial to ensure no one could have doctored them. They checked whether any of the experimenters was getting suspiciously high hit rates. They could find no hidden factors. Whilst their 30% hit rate is not as high as the 33% found by Tressoldi and Storm in their meta-analysis, as Watt stated, their result is sufficient to support the precognition hypothesis.[109]

***

The evidence we've looked in this chapter seems to suggest that weird phenomena in the form of telepathy and precognition might be real and can be measured even under well-controlled conditions. This supports the idea that consciousness can transcend space and time. If this really is the case, it is difficult to envisage how this can occur if consciousness is an epiphenomenon of the brain, as suggested by brain-based theories (Chapter 2). Just as with the weird phenomena of out-of-body experiences and reincarnation that we looked at in the last chapter, this is further cause to question claims that consciousness is entirely brain-based. In the next chapter we'll examine the suggestion that consciousness may not only be able to exist beyond the body and reach beyond space and time, but perhaps may also be capable of directly influencing the physical world through the power of the mind.

Chapter 5

Can Consciousness Directly Influence Matter?

If it's possible for the mind to function beyond the body and to transcend space and time as we've discussed in the last two chapters, might it also be possible for consciousness to influence our seemingly solid physical world? Like causing a wine glass inexplicably to slide across a table in front of witnesses without touching it, something someone I once knew swore they had made happen, although they were never able to repeat it despite many attempts. Stories of such weird feats of mind-over-matter, known as psychokinesis (or sometimes telekinesis), have been told throughout the ages. This is particularly so within religious traditions, with examples including the would-be Buddha levitating a knot of his hair as a symbol of his coming enlightenment, Moses parting the Red Sea to allow the Israelites to escape the chasing Egyptians and the ancient Greek mystic Apollonius of Tyana teleporting himself between cities.

Mind Over Matter

When I was growing up in the 1970s, one of the most famous people in the world was Uri Geller. Geller boasted a range of remarkable psychic feats, but what he became best known for was bending spoons using—he said—just the power of his mind (together with a little gentle rubbing from his fingertips). He seemed to be everywhere on TV and hung out with many of the rich and famous of the day. What was special about Geller was his assertion that what he was doing was genuinely paranormal and not simply a series of magic tricks.

Geller was keen to be subjected to scrutiny from scientists to confirm the veracity of his claims. There was a laboratory-based investigation of his supposed ability to discern the contents of drawings sealed from view in opaque envelopes. This study was conducted by Russell Targ and Harold Puthoff (who led a CIA programme to train psychic spies, known as Star Gate) and was published in the leading scientific journal *Nature*.[110] Targ and Puthoff reported that Geller was able to reproduce the contents of these hidden drawings at stunning levels of accuracy despite various controls put in place aimed at preventing Geller from cheating, such as isolating him in an electrically-shielded cage. As to his supposed powers of psychokinesis, another group of scientists, including the eminent quantum physicist David Bohm, conducted tests of Geller's metal-bending and concluded that they were 'very nearly' at the point where there is "no room for reasonable doubt that some new [i.e., paranormal] process is involved here". Whilst they retained a note of caution that it was possible that Geller was using conjuring techniques that they had not been able to detect, they nevertheless reported that their analysis of bent spoons showed that "no conventional physical or chemical explanation… is readily apparent".[111]

But not everyone was impressed. Uri Geller's most strident critic was the magician James Randi (now deceased) who, according to his website, was "the world's most tireless investigator and demystifier of paranormal and pseudoscientific claims… [and] a thorn in the sides of those who try to pull the wool over the public's eyes in the name of the supernatural".[112] Randi was convinced that Geller was a fraud. He made a number of allegations which—if true—would fatally undermine the findings of experiments conducted by the scientists who tested Geller's abilities, suggesting, for example, that Geller had an accomplice in the laboratory. Randi also staged demonstrations showing how Geller's spoon-bending could be replicated by magicians who were able to fool some scientists into believing that psychokinesis—rather than something more prosaic—was at work. He also pointed to video evidence that apparently showed Geller surreptitiously bending spoons with brute force in a distinctly non-paranormal manner.[113]

Incidents where objects are reported to have moved of their own accord or have supposedly been manipulated as a result of mental intention are also potentially subject to flaky memory and exaggeration, just like the other weird phenomena we've looked at. Chris French suggests that reports of psychokinesis may be additionally susceptible to what is known as 'inattentional blindness', an effect where we fail to notice things in our environment because we're concentrating on other things. This can lead us to think we see something that is not really there or miss something that is.

Skilled magicians take advantage of such effects to create compelling illusions using techniques of misdirection. This can be combined with a tendency to be open to suggestions from the magician or others in the audience that can fool us into believing we are seeing something that's not really occurring. As psychologist Richard Wiseman puts it, magicians "need to know how to make you misperceive what is happening inches from your nose, prevent you from thinking about certain solutions to tricks, and persuade you to misremember what has happened right in front of your eyes".[114] French suggests that many of Uri Geller's metal-bending feats rely on such techniques, for example when some people (including scientists) reported that they saw metal objects continue to flex even after being placed on a table. This may have been due to them responding to Geller saying something like "Look! It's still bending!"

The potential for Geller and others to trick an audience demonstrates the reputational risk for scientists who work with psychics claiming to produce dramatic large-scale effects. Whilst there are some parapsychologists who still perform investigations of what is known as macro-psychokinesis or macro-PK, most of those who are interested in this phenomenon have turned their attention to experiments that can more easily be protected from the possibility of cheating and typically involve ordinary people, who, unlike Geller, have little motivation to influence results by nefarious means. These experiments involve what is known as micro-PK, a term coined for how the mind might affect small-scale random physical processes as opposed to everyday objects.

During the 1930s and 1940s, J.B. Rhine, whose work on telepathy we encountered in the last chapter, carried out what were probably the earliest formal experiments of micro-PK. Rhine looked at whether it was possible to affect the roll of dice using the mind to achieve higher scores than we might expect by chance. He initially tested 25 people and found that they could indeed do this at a rate that was highly unlikely to be random, albeit with very small effects—over a large number of trials this averaged out at just 5.53 times in every 12 (46%) against the expected 5 in 12 (42%). Rhine realised that the dice themselves might be biased and more likely to show higher numbers because of the holes used for the dots, so he repeated the experiment except this time participants aimed for low, instead of high, scores. Again he found a statistically significant effect,[i] again at just a tiny rate. In 1990, Dean Radin and Diane Ferrari analysed 148 dice-rolling studies that were similar to Rhine's that had been conducted by 39 different scientists and found that they collectively achieved results that should only occur by chance once in 10^{70} times (although, as we discussed in the last chapter, there is reason to be wary of such meta-analyses).[115]

In 1978, Robert Jahn, then a professor of engineering at Princeton University, was approached by a student who asked him to supervise a project to design an electronic random number generator (RNG) that might be affected by human intention.[116] Whilst Jahn was doubtful that this was possible, he agreed anyway because he felt that there would be useful engineering involved and hence a good learning experience for the student. When to his surprise the student achieved positive results, he ran the tests himself and got similar results. This led him to establish the Princeton Engineering Anomalies Research (PEAR) laboratory to host an extended programme of research into micro-PK, which ran from 1979 to 2007. Among other techniques to test mental influence over physical processes, the team at PEAR built sophisticated RNGs that take advantage of fundamentally random effects such as radioactive decay or noise from electronic circuitry to generate a random stream of ones and zeroes—like repeatedly flipping a coin for heads or tails—and asked people to will the machine to produce more of one than the other, which they were consistently able to do. This type of experiment proved popular among other researchers and there have been several analyses to gauge the combined effect of almost a thousand separate studies with RNGs conducted at many different labs around the world. Whilst typically results show only a miniscule effect—less than a one percent difference between what is expected from true randomness and what

---

i A result that is statistically significant is unlikely to have occurred by chance and points to the likelihood that the effect in question is genuine (assuming that the study as a whole is sound). Although thresholds differ between projects and scientific disciplines, in psychology, 'significant' usually means that a similar result would only have occurred by chance less than one in twenty times. A result that is 'highly significant' would only have occurred by chance less than one in a hundred times. We'll encounter this meaning of the word in similar contexts throughout this chapter and again in Chapter 13.

people manage to achieve through mental influence—nevertheless, this effect is astronomically unlikely to occur simply by chance.[117] Although again, we need to treat such combined results with caution.

Collective Consciousness

Roger Nelson, one of the scientists involved in the PEAR experiments, was intrigued by the idea that human minds might be part of a universal consciousness. Such a concept has been proposed by the philosopher Pierre Teilhard de Chardin who suggested that the Earth might have a collective mind known as a noosphere (a name chosen for its conceptual similarity to the geosphere and biosphere), and by Carl Jung who proposed the idea of a collective unconscious. Nelson got to wondering whether RNGs might be susceptible to influence when there is a coherence of consciousness created by large numbers of people, for example when there are major events that attract mass attention. In 1997, Diana, Princess of Wales, world-famous for her humanitarian work, glamorous lifestyle and high-profile criticisms of the British royal family, died suddenly in a car crash in Paris after attempting to escape pursuing paparazzi. Because Nelson knew that her televised funeral was likely to attract a huge global audience, he asked colleagues to run their RNGs throughout the ceremony to test his idea. When he analysed the combined data from the RNGs, he found highly significant deviations from that expected by chance. He hypothesised that this was due to some kind of field effect caused by an unusually coherent human collective consciousness during the funeral that affected the miniscule quantum events measured by the RNGs.[118]

These encouraging results led Nelson and his team to establish tens of computers in various different countries hosting RNGs that would run continuously. The idea was to see whether unexpected notable public events would 'cause' (as the hypothesis would have it) significant deviations from chance in the data streams. In a nod to how an electroencephalogram (EEG) measures a person's brain waves, Nelson termed this network of computers that he suggested might be measuring the state of global human consciousness an ElectroGaiaGram (EGG), although the project is more formally known as the Global Consciousness Project (GCP). Since the project began in 1998, many globally-significant events have been associated with significant deviations in the random data stream. These include terrorist incidents like the September 11 attack on New York in 2001, natural disasters such as the Pakistan earthquake of 2005 and mass meditations such as the Jain World Navkar Mantra Day in 2025, which involved over one million participants in over one hundred countries.[119,120] A combined analysis of more than 500 separate events over two decades shows odds against chance in excess of a trillion to one.[121] There has also been analysis by Ulf Holmberg that suggests that the EGG might even anticipate before time the movement of global stock markets.[122] (We'll return to the GCP in Chapter 13 when we look at testing aspects of consciousness.)

A more recent initiative along complementary lines has been established by Vasilios Basios and Wolfhardt Janu. This adopts a slightly different approach to the GCP by examining patterns and correlations between the outputs of two parallel RNGs running on the same computer rather than just the one. Basios and Janu suggest that this can provide more sensitive readings and perhaps pick up on subtleties that might be missed from only a single data stream.[123] They and colleagues are particularly interested in potential field effects from more constrained events that might arise from fields of local (rather than global) consciousness. For example, intriguing early findings seem to show significant correlations between non-random patterns in the data and key stages within group tantric meditations by advanced practitioners. Another study involved placing these double-RNG machines in rooms within hospital intensive care units where someone is about to experience their end of life (with permission from those involved of course). This seems to show that a 'spike' consistently occurs in the data, not only at the exact moment of death but also—somewhat unexpectedly—two hours after the person in question has passed on.[124]

This research has given birth to a company called Wyrd Technologies, established by entrepreneur and author Peter Merry, which offers commercial products incorporating technology based on the principles established by Basios and Janu. This includes machines for researchers to use in experiments and lamps with random RNG data that are said to respond with different colours and intensities depending on the intention, mood and emotions of those present. Wyrd Technologies aims for these and similar technologies to make apparent to a wide range of people the reality of how consciousness might affect the physical world.[125]

Energy Healing

If consciousness can be shown to affect inanimate physical objects such as metal, dice or random number generators, perhaps it can also have an effect on living organisms. This specific type of mind-over-matter is sometimes described as 'biological psychokinesis'. Such focused conscious intention on living beings usually has the purpose of invoking beneficial therapeutic effects and so is often known as energy healing. Examples of energy healing include faith healing, practised in some Christian churches, which involves prayers for the intervention of God to cure the sick, and reiki, a Japanese technique that aims to work through 'subtle life force energy'. The sceptic Chris French states categorically that "all forms of complementary and alternative medicine [including energy healing]... have either never been proved to work or they have been proved not to work".[126] But is this really the case?

One study led by Leonard Leibovici that was published in the *British Medical Journal* looked at the effect of prayer on people who had been admitted to hospital with a bloodstream infection. The study was a randomised controlled trial, a methodology seen as the gold-standard for assessing the efficacy of medical interventions. More than three thousand patients took part and were randomly allocated to one of two groups—those who received a prayer for their well-being and full recovery and the other—a control group—who received no intervention of any kind. Neither the patients themselves nor the medical staff involved were informed as to which group each belonged. The study found that the group who were prayed for remained in hospital for a significantly shorter time than the group who didn't receive prayer. Their fever also broke sooner. Both of these results were statistically significant, meaning that they were unlikely to be a result of chance differences between the groups. It was also the case that fewer of the group being prayed for died in hospital than the control group (28% compared with 30%).[127]

Whilst this feels like an impressive result, there is an issue with these and similar studies looking at prayer and mental intention. This is that it is hard to isolate any potential effects of distant healing that happen within the experiment and the likelihood that patients in the study will be receiving prayers and positive thoughts from family and friends who have nothing to do with the experiment. This includes those in the control group who, according to the experiment, are not supposed to be receiving treatment. This can muddy the waters because, even if such intentions work, who can say whether improvements in patients' health are from the energy healing under investigation or from the good wishes of those close to them?

Irritable Bowel Syndrome (IBS) and Inflammatory Bowel Disease (IBD) are conditions that result in much misery for sufferers with symptoms that include abdominal pain, diarrhoea and constipation. In a randomised controlled trial led by R.T. Lee and carried out at two NHS hospitals in the Birmingham area of the UK, 200 patients with IBS or IBD were randomly allocated either to a control group, which was given conventional treatments for their condition, or to an intervention group, which was given spiritual healing in addition to conventional treatments. The spiritual healing consisted of weekly 30-minute sessions involving light touch from a trained therapist. A highly significant improvement in the severity of both physical and mental symptoms was reported by those in the intervention group compared to the control group. The control group was given spiritual healing 12 weeks into the study at which point they also reported similar improvements. The authors note that the "benefit observed may at least be, in part, a placebo effect although the size of benefits observed suggests an alternative mechanism and the value of any mechanism should arguably not be discounted where it confers symptomatic relief".[128]

It can be difficult to disentangle the effects of energy healing in experiments with human participants from other factors. We are complicated beings. For this reason, scientists have also looked at the effects of energy healing that don't involve human participants and so should not be subject to influence outside of the experiment. One example of this type of research led by Chang-Tze Yu looked at whether water treated with positive intentions might help to arrest the activity of cancer cells in a laboratory petri dish.[129] Four Buddhist monks were asked to 'imprint' the following intention on some bottles of ultrapure water: "the human cancer cells that absorb this water will obtain beneficial changes or alleviate malignant properties". A control set of bottles of ultrapure water that was otherwise the same was left unimprinted with any intentions. Each set of water was applied to a separate line of cancer cells taken from a glioblastoma tumour, a particularly aggressive form of brain cancer that is exceedingly difficult to treat. Individuals treating the cancer cells were unaware of whether the bottles of water they were using had received the monks' good intentions or whether they were the control bottles. This meant that any unconscious behaviour on their part couldn't inadvertently affect the results. The cells that received the positive intention-treated water were found to behave in a significantly less virulent manner than the cells that received the untreated water. The authors concluded their paper with a recommendation that clinical trials be held for this treatment involving actual brain cancer patients, something that would be the usual next step for more conventional treatments that display similarly promising results in the lab.

Unexceptional Evidence?

A pithy retort often made by sceptics to parapsychological findings is that "extraordinary claims require extraordinary evidence". Results such as those we've discussed in this and the previous two chapters are said to fall far short of this threshold and are therefore summarily dismissed. This statement is repeated so often that it has become known in some circles by the acronym ECREE. Whilst ECREE is usually attributed to the science communicator Carl Sagan, this principle, or something like it, has also been suggested by a number of historical figures including David Hume and Thomas Jefferson.[130]

There are two parts to ECREE, so let's examine each in turn. The first is that it is an extraordinary claim that weird phenomena are real. However, representative surveys have shown that around two-thirds of the population believes that at least one such phenomenon is genuine.[131] Whilst some might put this down to gullibility on the part of the general public, another survey found an equivalent level of belief even amongst highly educated scientists and engineers.[132] The claim that weird phenomena are genuine might therefore be seen as commonplace and perhaps not so extraordinary. What is beyond the ordinary is anyway rather a subjective judgement, which is a tricky notion to invoke when discussing matters of science.

The second part of ECREE is its demand for extraordinary evidence. The problem with this is that, whilst individual reported experiences of spontaneous occurrences of weird phenomena are indeed often extraordinary, well-designed parapsychological experiments that rule out alternative explanations typically provide merely ordinary evidence. Similar levels of evidence are regarded as sufficient in many other areas of science such as medicine and psychology. This can be seen by comparing 'effect sizes' found in experiments. As well as measuring how likely any effect found in a study might be down to chance, known as its 'statistical significance level', experiments also measure the size of this effect. This is useful in determining whether the phenomena being examined have real consequences in the world.

Average effect sizes that are found in experiments for telepathy are around 0.14, for precognition 0.2, for psychokinesis using RNGs 0.5, and for energy healing 0.2.[133] These are regarded as small or medium-sized effects. However, these values are larger than effect sizes found for other widely accepted psychological effects such as how people attribute their failures to being unlucky or how people who are seen as more credible are also more persuasive, both of which have effect sizes of around 0.1.[134] Weird phenomena also exhibit similar or larger effect sizes to those found for the effectiveness of common pharmaceuticals such as the use of antihistamine for sneezing and ibuprofen for pain relief, which have effect sizes of 0.11 and 0.14 respectively.[135] If we're prepared to take a pill for hay fever or a headache on this basis, why is a similar level of evidence for telepathy, precognition or psychokinesis regarded as insufficient just because sceptics deem it unextraordinary?

Perhaps it's because findings from parapsychology are viewed more sceptically than those from other disciplines even when all else is equal. This was demonstrated in a piece of research by Bethany Butzer where she randomly assigned 100 people with a background in psychology to one of two groups.[136] One group was asked to assess an evaluation of parapsychological research and the other an evaluation of neuroscientific research. The text each group was given was the same except for subject-specific references. For example, the first sentence of the parapsychological summary read "This article presents a comprehensive integration of current experimental evidence about parapsychological phenomena", whereas the neuroscience summary read "This article presents a comprehensive integration of current experimental evidence about the role of the hippocampus in storing episodic memories". Similar substitutions were made throughout so that the summary and evidence presented for each topic was identical. Despite the exact similarity of the two evaluations in terms of their substance, participants rated the parapsychological evaluation as being weaker and less reliable than the neuroscience evaluation, showing an effect of confirmation bias or even straightforward prejudice against parapsychology.

Some sceptics account for positive experimental results by suggesting that the standard of parapsychological research is simply not up to scratch. They argue that any positive experimental findings must be due to shoddy methodology, dodgy statistics, reporting only studies that show an effect or even outright fraud. Some even go so far as to suggest that parapsychology is not really science at all, that it is pseudoscience, on a par with creationism, climate change denial and the Bermuda triangle.[137] This last claim is particularly pernicious because it is all over *Wikipedia,* the entries of which are shown prominently in internet search results and on electronic devices such as some e-readers. Such content is created—and jealously guarded—by organised teams of editors with somewhat dogmatic views of parapsychology who describe themselves as 'Guerilla Skeptics'.[138] Many people are not aware of the source of this one-sided material and might therefore believe that this is a reliable point of view, something that is disputed by a number of commentators.[139]

Whilst of course not all parapsychological experiments are exemplary in design and/or execution, there is no real evidence to suggest that parapsychology suffers from lower standards of research than any other field. Indeed, history has shown that scientists investigating weird phenomena have repeatedly implemented improvements to methods and tightened up procedures in response to valid criticisms of their work. Such criticism is part and parcel of good science—it is how standards improve and findings become more robust. For example, J.B. Rhine built a mechanical dice-throwing machine for his psychokinesis experiments in response to suggestions that people might be able to attain certain scores by how they threw the dice rather than by mental intention. In modern parapsychological experiments, it is routine to implement designs that incorporate strict controls and safeguards to eliminate as far as possible explanations other than the paranormal. And indeed, not all sceptics suggest that parapsychology is pseudoscientific. For example, Chris French has this to say:

> In deciding whether parapsychology is better described as a science or a pseudoscience, we should look at the methods employed by parapsychologists. It would be unfair to judge any discipline only by considering the poorest work done within it (if we did so, psychology would certainly be dismissed as a pseudoscience). Parapsychology at its best... appears to meet most if not all of the benchmarks of true science as opposed to pseudoscience. That is so even if paranormal forces do not actually exist.[140]

Unfortunately, some sceptics have decided that it is not worth even looking at parapsychological evidence. In 2018 one of mainstream psychology's leading scientific journals published an article by Etzel Cardeña suggesting that the weight of experimental evidence amounted to 'cumulative support' for certain weird phenomena.[141] Cardeña's review summarised more than 2,500 parapsychological experiments that focused on ESP and psychokinesis that have been conducted over the last 90 or so years and, he suggested, collectively indicate the reality of these phenomena. Cardeña's conclusion was that this means that such weird phenomena are genuine and should be taken seriously.

A subsequent issue of the same journal and an article in a sceptical publication featured a forthright response and rebuttal of Cardeña's paper from psychologists Arthur Reber and James Alcock.[142,143] Reber and Alcock stated that any evidence showing that weird phenomena might be real simply "cannot be true" because "pigs cannot fly". Their reasoning was that "paranormal effects violate basic scientific principles" meaning that experiments that produce evidence that supposedly support such weird phenomena must necessarily be 'flawed'. Their confidence in this line of argument led them to assert that they "did not examine [Cardeña's] data... Our reason was simple: the data are irrelevant".

***

The research discussed in this chapter offers support for the view that weird phenomena involving mind-over-matter in the form of psychokinesis, field effects and energy healing are not only real but that their effects can be demonstrated in well-controlled scientific experiments. Findings from studies seem to suggest that consciousness can directly affect physical matter. Like the weird phenomena showing that consciousness might exist beyond the brain and body (Chapter 3) and that consciousness can transcend space and time (Chapter 4), the various types of mind-over-matter we've looked at in this chapter once again challenge the foundations of those suggesting that consciousness is solely a product of the brain. Whilst sceptical explanations are sometimes sufficient to account for some reports of weird phenomena, it is harder to discount them all due to the wealth of experimental evidence supporting their existence. But there is one other phenomenon beyond those already discussed that those who support a brain-based view of consciousness also typically reject although most of us don't normally consider this as at all weird. This is the notion that we have free will.

Chapter 6

Do We Have Free Will?

Most of us live our lives with the strong sense that we exercise conscious choices in the world. We feel that for every decision we make, we could have made another. For instance, it seems to me in this moment that, if I so choose, I could stop typing this sentence and go and make a mug of coffee. Or a cup of tea. Or dance around the room. Or indeed any number of things. It's true that the choices we make and actions we take are to some extent influenced by our personality, current situation and immediate environment, and by what is physically possible, but that still leaves a lot of options open. It seems that we are largely free to exercise our will and the vast majority of us live our lives on this basis.

Free Will Contradicts Physical Laws

If we subscribe to the view that consciousness arises solely from brain processes, as the theories we looked at in Chapter 2 would have it, then our minds are entirely the result of the biology and chemistry of the brain. From this perspective, free will is just another weird phenomenon, albeit one that is taken for granted almost all of the time by almost all of us. Brain-based theories imply that our consciousness is ultimately reducible to physical processes that are wholly deterministic. There is simply no mechanism in such theories for us to have any will to exercise in the world. Our impression of being free must therefore be the result of an illusion, albeit an immensely powerful one. We are entirely at the mercy of neural processes, which collectively determine every action we take. We are not free at all.

The root of this argument is that everything that happens has a physical cause, which in turn has something that caused it. In his book *Determined: Life Without Free Will*, biologist Robert Sapolsky argues that all of our behaviours are caused by the activity of neurons. These are stimulated by thoughts, memories, emotions and inputs from our senses. These manifest according to the way our brain was formed in childhood, which is due to our upbringing and our genes. Our upbringing is influenced by the culture in which we were raised, which itself developed through how generations that came before us lived. Our genes evolved through pressures on our species related to survival. These pressures are due to biological, environmental and ultimately physical factors. And on and on. In other words, writes Sapolsky, it's "causes all the way down".[144]

Another strong advocate for this view is the physicist and popular YouTuber Sabine Hossenfelder, who maintains that because brains are made of particles and because we know how particles behave through the laws of physics then we can fully explain how brain processes work. Which leaves no room for free will. Whilst we may not be able to predict with certainty what will happen next because of the complexity of the brain and its interactions with the body and the outside world, nevertheless, she says, the principle holds that the "future is determined by the present" with no role for consciously-willed actions. Anyone who disputes this, she says, is "contradicting well-established science".[145]

Sam Harris, a neuroscientist and philosopher, also argues that the universe works entirely through cause and effect. He suggests that free will is "worse than an illusion; it's actually a totally incoherent idea". Like Sapolsky and Hossenfelder, he suggests that what is happening in any particular moment depends entirely on what happened previously. Hence what we think and what we do is entirely the result of our current brain-state, which, in turn, is the product of the genetic make-up we inherited from our parents combined with our life experiences. None of these are in our control. Hence we have no free will.

Harris maintains that the logical conclusion of this deterministic view leaving no room for free will is that we can bear no responsibility for our actions, whether these are for the good or are downright evil. This has major implications for our justice system because it means that punishment for crimes cannot be justified. For example, he says:

> Consider a generic serial killer. His choice to commit his last murder was determined by neurophysiological events in his brain which were in turn determined by prior causes, bad genes, the developmental effects of an unhappy childhood, a night of lost sleep because a car

> alarm was going off... As sickening as I might find such a person's behaviour, I have to admit that if I were to trade places with him atom for atom, I would be him. There's no extra part of me that could resist the impulse to victimise innocent people.[146]

Others offer a more nuanced perspective. The physicist Jim Al-Khalili questions the forthright assertions often made by those who view the universe as fundamentally deterministic. Whilst he argues that the universe is ultimately based on cause and effect, it is also in practice fundamentally unpredictable, an idea expressed through chaos theory. Chaos theory describes how very small changes in a complex system can lead to very large differences in outcome, a phenomenon known as the 'butterfly effect' because of how the tiny impact of a butterfly flapping its wings in a rainforest in the Amazon can lead to diametrically opposing outcomes, such as making the difference to whether it's sunny or it's raining in the UK some weeks later. Hence the result of any cause is impossible to predict. We can't know what any of us might choose to do next because of the complexity of our brains and this is what gives some succour to the concept of free will, or at least our perception of it. He writes that "whether we call [this] true freedom or just an illusion in a way does not matter".[147]

Some who support a brain-based view of consciousness suggest that there is room for free will but only by framing the issue somewhat differently to how most people conceive of it. One such is philosopher Daniel Dennett who argues that, despite us living in a fundamentally deterministic universe, we should hold on to the idea of free will because it is the basis for morality, which underpins the way we live and organise society. He suggests that whether events and processes are ultimately deterministic is not relevant to questions of right or wrong and hence it is worth preserving our intuitive sense that we have freedom of choice.[148] This type of position is known as 'compatibilism' because it suggests that free will is after all compatible with determinism, albeit by seemingly redefining free will to be something other than what most people would probably consider it to be.[149]

Neuroscientist Anil Seth (who spends much of his working life conducting research into the predictive processing model of a brain-based consciousness discussed in Chapter 2) agrees that we do not have free will in the sense of being able to make meaningful decisions in the world. Like the authors discussed above, he also believes that our minds cannot meaningfully choose between alternative actions. In any instant we are constrained to act in a particular way because of our circumstances, our personal and cultural history and, ultimately, our brain's circuitry, meaning that it is simply not possible that we might have acted in a different way. What Seth offers is an explanation for why it is that we retain a strong feeling of free will despite these constraints. Seth suggests that the subjective experience of free will is simply how we experience the brain operating in the way that it does. So why is that we feel like we could make a decision other than the one we actually take? He responds:

> Well, it matters for the next time. The brain pays attention to voluntary actions because the brain is always changing. Maybe it turns out that I've suddenly become allergic to tea and I have this cup of tea and I feel terrible. So now the brain has learned that, okay, next time coffee will be made rather than tea or water or something else, because the alternative possibilities have been highlighted in my experience the first time.[150]

Brain Processes Happen Prior to Decisions

One of the most frequent lines of argument that there is no such thing as genuine free will stems from the findings of experiments led by the neuroscientist Benjamin Libet. His results appear to show that the unconscious brain processes that underpin any particular action occur definitively *before* a conscious decision is made to take that action.[151] This means that when we make what we believe to be a free choice this decision in fact takes place *after* neural processes happen that actually determine this choice. A few hundred milliseconds separates the unconscious 'decision' made within the brain from the subsequent conscious subjective 'decision' that we perceive (wrongly) as freely made. This seems to confirm the brain-based view of consciousness that we have no free will. This also appears to accord with Seth's suggestion that free will is essentially just the term we use to describe the experience of certain brain processes in circumstances that appear to involve a free choice. Sam Harris has said in light of these experiments that it is therefore

> scientifically uncontroversial to say that some moments before you were aware of what you are going to do in making a simple voluntary action… your brain has already determined what it is you will do and then you become gradually aware of this decision.[152]

Similar observations have been made in relation to other cognitive processes that seem like they occur without pre-thought, such as speech and superfast reactions in sport. These examples may suggest that what we feel to be free choices are instead the effects—and not the causes—of processes in our brains relating to these 'decisions'. Unconscious neural processes are determining what we think and do and we ourselves have no decision-making role.

Some of the theorists we will meet in the next five chapters have addressed Libet's experimental findings in terms of their various perspectives that consciousness is not entirely explicable through brain-based mechanistic processes. Libet's results might also potentially be explained through some form of presentiment effect similar to those discussed in Chapter 4, as has been noted by my colleague Piero Calvi Parisetti,[153] although the implication of unconscious brain processes anticipating what a person is about to do could have just as significant implications for free will as does a deterministic view.

However, there are reasons for questioning the standard interpretation of Libet's experiment without invoking theories of consciousness extending beyond the brain. Experiments similar to Libet's that have been performed by neuroscientists such as Christoph Herrmann and colleagues suggest that what Libet may have been measuring in his participants' brains was not some unconscious processes leading deterministically to a decision but instead the brain's 'readiness potential'.[154,155] These researchers note that participants in Libet's experiment were told to expect that they would need to perform some movement and so all that Libet might have been measuring was the neuronal processes in the brain preparing for such a movement rather than the choice itself. As Herrmann and colleagues write, the neural activity often cited as an observation of the brain predetermining a course of action "instead... may reflect a general expectation. This interpretation would not interfere with the notion of free will".[156] This means that conclusions from this oft-cited study may not be quite so clearcut in terms of their supposed implications for free will.

Does Quantum Mechanics Rescue Free Will?

What is the response to the view that free will does not exist, at least in the sense of us having genuine agency in the world? That free will, like other weird phenomena, might be just what it appears to be? We will examine how various theories of an extended consciousness address this question over the course of the next five chapters; however, it is worth mentioning here that many defenders of free will invoke quantum mechanics, one of the foundational theories of modern physics. In the next chapter, I'll provide a short summary of some aspects of quantum theory, but suffice to say here that quantum mechanics blows a hole in the idea that the world is entirely deterministic, that causes always precede effects, at least as regards the world of the very small.

Whilst the authors we've so far discussed are of course fully aware of this dimension of quantum mechanics and its inconsistency with the idea of an entirely mechanistic universe, they dismiss the notion that this can provide any explanation for free will. They point out that the nature of quantum mechanics is that it is essentially random. They maintain that quantum mechanics does not allow for free will because we have no control over random events, just as we have no control over events that have been determined by previous events. As Sabine Hossenfelder says, "quantum events are not influenced by you... because they are not influenced by anything. That's the whole point of saying they are fundamentally random".[157]

Nevertheless, others disagree and suggest that quantum mechanics might indeed allow for free will. The arguments against free will rely on a deterministic universe. However, as physicist Ruth Kastner notes, quantum mechanics implies that the future is not determined by previous events. This is a prerequisite for free will to be genuine. Instead, she argues that there exist in the world described by quantum mechanics what she calls 'unactualized', or unrealised, possibilities.[158] But free will implies more than just an indeterminate world. What is also required is for us to have genuine freedom to be able to exert some control over what may happen. Kastner suggests that this can be accounted for by the twin nature of reality that is described by both classical physics, which is deterministic, and quantum mechanics, which is not. She suggests that minds and thoughts exist outside of spacetime in the quantum realm, and so they are therefore not subject to classical physics. She writes that the

> predictability of classical physical laws emerges only at the macroscopic level of actualized spacetime events. At the quantum level, the world is genuinely indeterministic. And this is where our free choices can enter... [These are] at different levels of reality, so that they do not have to conflict with each other.[159]

This is just a taster of some of the arguments we'll encounter over the next five chapters that suggest that consciousness is not entirely due to a mechanistic brain and deterministic neural processes.

A Logical Paradox

Before we leave this topic, it is worth noting the often paradoxical nature of arguments expressed by those who believe that free will is essentially illusory. For example, Sam Harris makes the case for reforming the justice system to account for people's lack of responsibility for their own predetermined actions. He argues that punishment should be abolished because it is predetermined that criminals will commit crime. They have no choice. The paradox comes when considering how we might go about bringing change to the justice system—or indeed any system—when, it is argued, we have no free will with which to do so. Authors such as Harris appear unwilling to accept the fatalism that would seem to be the logical conclusion of arguing that we have no agency in the world.

Harris suggests that, despite free will being 'worse than an illusion' and 'incoherent', it would be a mistake to conclude that choices therefore do not matter and that it is not possible to change. He suggests that self-improvement is still possible without free will: "clearly, choice and effort are part of the causal chain of life. If I hadn't decided to write a book on free will it wouldn't have written itself."[160] I find statements such as this from Harris and others somewhat startling considering their contention that everything is determined and we ultimately have no control in the world. It is hard to reconcile a viewpoint that regards choices as important and meaningful whilst at the same time being entirely illusory. As the author of *Determined*, Robert Sapolsky, puts it, "even I think it's crazy to take seriously all the implications of there being no free will".[161]

***

So we come to the end of our examination of weird phenomena. Perhaps the evidence I've presented over the last four chapters has been sufficient to convince you that it is possible for consciousness to exist beyond the brain, to transcend space and time, to directly influence matter and that we might have genuine free will. Or perhaps you have more sympathy with the sceptical view that argues that all of these apparent weird phenomena can be adequately explained by normal psychological or neurological effects without resorting to the paranormal or accepting that free will is genuine.

It is worth pointing out that it is sufficient for just one weird experience to be what it appears to be for there to open up a fatal flaw in the assumption of brain-based theories that consciousness is confined entirely within our skulls and is simply the result of brain processes. As the pioneering psychologist, and one of the leading thinkers of the nineteenth century, William James, wrote:

> A universal proposition can be made untrue by a particular instance. If you wish to upset the law that all crows are black, you mustn't seek to prove that no crows are; it is enough to prove one single crow to be white.[162]

So now we come to look at theories of consciousness that appear, in principle at least, to be able to accommodate the possibility that at least some weird phenomena are genuinely what they appear to be, that at least some crows are white. These theories are those that suggest that consciousness can extend beyond the brain.

Chapter 7

Consciousness is Quantum

If we accept the evidence for weird phenomena then it seems that brain-based theories cannot provide a complete and satisfying explanation for consciousness. If our minds really are just the product of a kilogram or so of mushy brain matter within the skull then it is difficult to imagine how that consciousness can breach such physical constraints to continue beyond death, extend beyond space and time, influence matter and shape its own destiny. So what about other explanations that look beyond the brain? Might these more plausibly account for this more expansive view of our minds?

Over the next five chapters, we'll examine ten candidate theories that could—at least in principle—more ably accommodate at least some of these weird features of consciousness. These theories take a number of perspectives in suggesting that consciousness is essentially quantum mechanical, that it is hyperdimensional, field-like, informational or that it is fundamentally all there is. We'll start with the first of these categories—those that suggest that consciousness is in some way related to the world described by quantum physics.

Before we look at two of these approaches in more detail, we need to understand a bit about quantum mechanics. This theory, which we touched on in the last chapter, was first proposed in 1925 by physicists including Erwin Schrödinger, Werner Heisenberg, Max Planck, Albert Einstein, Niels Bohr and Louis de Broglie. Quantum mechanics is our most fundamental theory of how the physical world works. In many descriptions of this theory, it is common to refer to many of its insights about reality as odd, strange or, indeed, weird. This is because quantum mechanics contradicts what has become known as the classical view of the world that held sway until the early twentieth century and, as we saw particularly in our discussion of free will in

Chapter 6, still dominates how many people think the universe works. In classical physics, the universe is seen to operate like a machine where each effect has a cause and everything is in principle determined. In this schema, events are, in theory at least, entirely predictable from a precise knowledge of all of the starting positions involved. However, quantum mechanics shows us that something radically different is going on and that the world doesn't work this way. Its premise differs from classical physics and makes surprising and seemingly counterintuitive predictions that have nevertheless been experimentally confirmed to enormous levels of precision.

Let's take a look at three of the key features of quantum mechanics that are most salient to our discussion of consciousness in this chapter.[163] The first of these is the principle of non-determinism. Quantum mechanics shows that it is impossible to predict exactly the behaviour of any subatomic particle—the building blocks of our physical reality—such as an electron or a photon. Any particle's properties, for example in terms of its position and how it's travelling, can never be fully known except in terms of likelihood or probability. Subatomic events are fundamentally uncertain until they are measured or observed, a situation often described as a collapse of a probabilistic wave function into a definite point-like particle. This wave function is not a tangible object existing in space and time. Instead it is essentially a statistical entity with no physical form suggesting that reality is no more than a mathematical description of probabilities until we measure it, rather than anything solid.

This leads us to the second principle, that of superposition. Prior to the collapse of the wave function, we might think of a particle as being in an indeterminate, spread-out state. Until we take a measurement, a subatomic particle can quite literally be in multiple different places or have multiple different velocities at once, despite this being seemingly impossible in commonsense terms. This has been compared to a coin that is simultaneously both heads up and tails up. The physicist Barton Zwiebach has commented that superposition is "a ghostly, eerie state" where particles do "incompatible things simultaneously".[164]

The third principle is known as entanglement. This is where two particles can appear to be inextricably connected despite their being physically separated even across immense distances. If one of two entangled but separate subatomic particles is observed to have a certain characteristic then its entangled partner will instantaneously take on this same characteristic without any information passing between the two. This happens no matter how far apart the particles may be. Even if they are at opposite ends of the galaxy these separated particles remain inextricably entwined as if they were somehow part of the same system.

Albert Einstein contributed another adjective to our list of descriptors of how mind-bogglingly strange quantum mechanics is—he called entanglement with its seemingly faster-than-light communication 'spooky action at a distance'.[165] He used the word spooky in a derogatory sense to show that he believed that entanglement could not possibly be an actual feature of the real world because his theory of relativity holds that nothing can travel faster than light and that therefore there must be something wrong with the mathematics that underpins quantum mechanics. However, experiments have shown that entanglement is indeed a real phenomenon and not just theoretical.[166] Entanglement seems to suggest that the building blocks of our physical world are non-local, meaning that they're not restricted to the apparent constraints of space and time. This is one of the reasons some researchers believe that quantum mechanics may be related to a consciousness that is likewise not constrained by space and time.

For many decades, it was assumed that quantum mechanical effects could only take place at very small scales and so were not relevant to the human-sized world in which we live. It seemed safe to assume that all objects and processes in our everyday world—including the brain—operated along entirely classical and deterministic lines, and therefore none of the strange features of quantum mechanics outlined here was relevant to how consciousness works. This assumption is what underpins brain-based theories of consciousness (Chapter 2). However, there is now an increasing body of evidence to support the idea that quantum effects can and do take place in biological systems including brains. Examples of processes that have so far been discovered include vision and smell in animals, birds navigating to migrate over long distances and photosynthesis in plants, all of which involve elements of quantum mechanics.[167] With this in mind, some theorists have suggested that quantum mechanics may provide the key to understanding consciousness.

Consciousness is the Quantum Wave Function Collapsing in Microtubules

The first theory we'll look at in this section is proposed by the neuroscientist Stuart Hameroff, and mathematician and holder of a Nobel prize in physics Sir Roger Penrose. They propose that a single conscious experience in humans is equivalent to an orchestrated collapse of the quantum wave function across the brain from a state of superposition to a definite state. Our consciousness is a continuous sequence of these individual conscious quantum collapse moments happening within microtubules, tiny structures found within neurons.[168] Hameroff and Penrose note that this idea of consciousness being a constant stream of discrete thoughts resonates with teachings from some ancient Eastern philosophies, some branches of which have even quantified the number of thoughts that might take place in a 24-hour period.[i]

Most neuroscientific theories of consciousness, such as those discussed in Chapter 2, are based on the assumption that the brain is a computational system. This system is said to comprise a neuron as its most fundamental unit, and then, at ever larger scales, networks of neurons, networks of networks of neurons, until ultimately cognitive computation encompasses the whole brain and this is said to be what we experience as consciousness. However, in his 1989 book *The Emperor's New Mind*,[169] Penrose questioned whether this model could ever lead to a satisfactory theory of consciousness. He argued that certain conscious experiences cannot possibly be computational, in particular the conscious sensation we have when we come to a point of understanding. This is the 'aha' moment that comes from an experience such as getting a joke or solving a puzzle or a crossword clue. This means, therefore, that human consciousness cannot be algorithmic and hence cannot be accounted for by networks of neurons. No physical computational structure such as the brain—or, come to that, an electronic computer, no matter how complex—could ever give rise to consciousness.

Penrose's conjecture is based on Kurt Gödel's incompleteness theorem. Gödel was an influential twentieth-century mathematician and philosopher who showed using a sequence of logical steps that no computational or mathematical system can possibly be complete in

---

i In case you're wondering, some 6,480,000 thoughts per day according to the Buddhist order of Sarvaastivaadins.

itself. There will always be something that is true that lies outside of that system. Penrose argued that because the brain is essentially a computational system that uses a network of neurons, for certain conscious experiences to exist, such as moments of understanding, Gödel's theorem means these must exist in a form that is beyond the physical matter of the brain itself.

Penrose argued that quantum wave function collapse is a prime candidate for this non-computable process that could equate to consciousness. Together with Hameroff, he identified microtubules as the most likely site for such quantum processes to take place. Microtubules are lattice-like structures that have a number of biological functions such as providing shape for cells and playing a key role in cell division.[170] Neuroscience research shows us that microtubules also have potentially consciousness-related functions. For example, microtubules are much more common within neurons than other types of cells, they seem to be involved in thinking processes such as facilitating connections between synapses, and cognitive brain impairments that are associated with diseases such as Alzheimer's seem to involve abnormalities and reductions in the number of microtubules.[171]

Microtubules also have a particular structure that, as Penrose says, "reeks of something quantum mechanical" in that its tubular shape may help to shield a delicate wave function superposition from the random noise of its surroundings and so prevent its premature collapse.[172] So, instead of considering neurons as the fundamental basis of consciousness as do brain-based theories, Hameroff and Penrose suggest instead that we must venture further inward to still smaller scales, to reach inside of neurons to microtubules, which could play host to the quantum collapse of the wave function. This collapse equates to a conscious moment of experience when it occurs simultaneously across a number of sites in the brain in synchrony in an 'orchestrated' fashion. Invoking another quantum effect, Hameroff and Penrose suggest that this orchestration of whole or regional brain states takes place through quantum entanglements between microtubule sites across different parts of the brain. As Hameroff puts it, this is "a quantum system embedded in a classical system of microtubules".[173] In a sense, this orchestrated wave function collapse is the brain engaging in quantum computation through moving from a superposition of entangled potential states into a single definite state. Indeed, whilst Hameroff and Penrose do not believe that a human-like consciousness can arise in a conventional computer, it is possible to conceive that something similar to consciousness may occur in a reasonably complex quantum computer that involves orchestrated (entangled) wave function collapses.[174]

Hameroff and Penrose's theory is not exclusively brain-based because microtubules—and therefore this hypothesised wave function collapse—can be found in virtually all animal and plant cells. Hence this theory could explain the apparent signs of consciousness we see even in single-cell organisms that can display surprisingly complex behaviour such as finding food, accessing memories and being able to navigate mazes in laboratory experiments.[175,176] Such organisms, of course, having only one cell, do not possess the brain structures and brain processes that are proposed to underpin consciousness by the brain-based theories we looked at in Chapter 2. Another way this theory differs from solely brain-based theories is that it is not attempting to equate consciousness with some high-level mechanical brain process. It is instead a theory of non-local consciousness because of its suggestion that conscious experiences are essentially quantum effects, which are fundamentally non-local. This means that, as we'll see later in this chapter, this theory might be able to account for certain non-local 'weird' aspects of consciousness.

Because wave function collapses occur everywhere throughout the universe, and not just in cells, neurons and brains, Hameroff and Penrose speculate that this might mean that primitive types of conscious experiences (proto-consciousness) may likewise be ubiquitous. Hence, perhaps such primitive consciousness has a pivotal role in how the physical world works in much the same way as do known features of fundamental particles, such as mass, spin or charge. However, these instances of isolated proto-consciousness are not of the same order as our rich experiences of consciousness; they do not encompass information or meaning because they lack the orchestration that occurs in more complex structures like the brain. Hameroff and Penrose ultimately conclude that it could be reasonable to infer from their theory that consciousness might be a "spiritual quality of the universe" that is fundamental and separate from the purely physical aspects of the world.[177]

Consciousness Acts on the Brain Through Quantum Effects

Our second approach to consciousness from a quantum perspective is championed by the mathematical physicist Henry Stapp, building on theories developed by another mathematical physicist, John von Neumann.[178] Stapp's theory relies on a particular way of interpreting quantum mechanics that is known as the Copenhagen interpretation. Whilst the fundamental mathematics and features of quantum mechanics are not in dispute, there are many differing interpretations of what it all might mean about the way the world works. Before we discuss Stapp's ideas in detail, we need first to understand this interpretation of quantum mechanics. And this is easiest to comprehend if we take a look at a famous foundational experiment associated with quantum mechanics known as the double-slit experiment that may help to explain the rationale behind the Copenhagen interpretation.

The double-slit experiment involves a light source—usually a laser—that can fire particles of light (known as photons) at a screen. There is a barrier placed between the laser and the screen. This barrier has two openings in the form of parallel vertical slits. There are two ways to set up the experiment in its most basic form. In the first (Figure 5), the light passes through the slits and we look to see what happens on the screen. It turns out that this produces a pattern with multiple lines of shadow and light suggesting that the light is acting as a wave, splitting into two when it passes through the slits after which each wave interferes with each other. This leads us to believe that light travels in waves like ripples in water caused by dropping a stone into a calm pond. Indeed, we would see something similar if we set up the experiment with a water wave passing through two gaps in a barrier rather than light.

In the second set-up (Figure 6), we place a detector at the barrier so that we can tell which slit each individual photon passes through on its way to the screen. If you haven't come across the world of quantum physics before, and specifically this experiment, you might predict that we would get the same result as in the first set-up. After all, why would light behave differently simply because we've taken a look at it at a certain stage in its journey? But instead of an interference pattern, here we find that the screen shows just two vertical lines. We might see something similar if we set up the experiment with small solid balls being projected at two gaps in a wall instead of light.

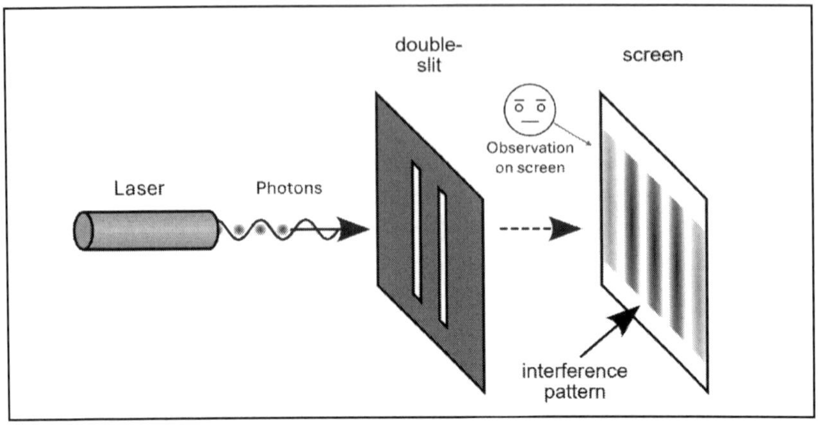

Figure 5. Illustration of the double-slit experiment with observation/measurement at the screen

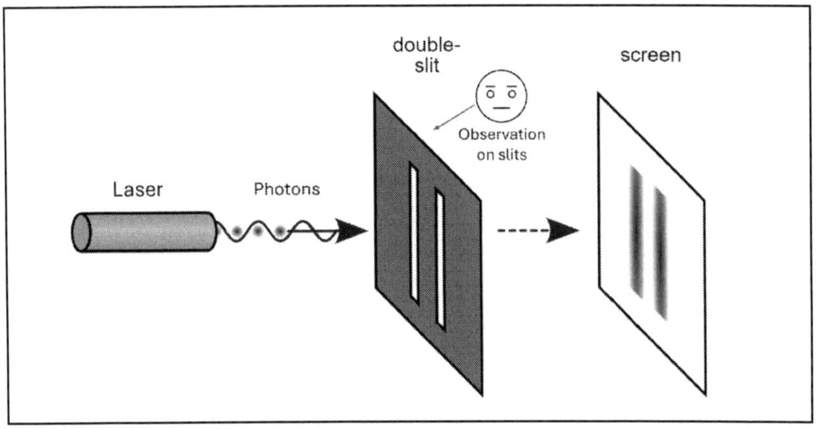

Figure 6. Illustration of the same double-slit experiment, this time with observation/measurement at the barrier

So is light a wave (like water waves or sound waves) or is it made from particles (like tiny solid objects)? And how can just making a measurement at the barrier seemingly determine how light behaves?

The most prominent and widely accepted interpretation among quantum physicists of what is going on here is the Copenhagen interpretation,[179] named after the city where many of the key features of quantum mechanics were first developed in the early twentieth century. This is the interpretation favoured by Stapp. The Copenhagen interpretation suggests that an observer of a quantum system is not some neutral bystander outside of this system but instead plays a central role in determining its behaviour. The very act of making an observation or taking a measurement of what is going on is what causes an indeterminate and spread out wave function to collapse into a definite point-like particle.[180] If we observe at the screen, the collapse happens here and light seems to behave like a spread out wave. However, if we observe at the barrier, the collapse takes place here instead and from this point on, light behaves like particles.

Exactly who or what this observer might be that collapses the wave function has often historically been left somewhat vague by those arguing for this interpretation. For example, is it sufficient for the observer to be an inanimate detector like the one used in the double-slit experiment, a mere piece of machinery? Or does it need to be a sentient organism? In which case, what counts as sentient? Might an insect, say, collapse the wave function? Or does collapse require a more complex conscious creature such as a cat or a human? As an attempt to resolve this question, Stapp and von Neumann suggest that the observer in any quantum system must be consciousness itself.[181] For example, in a quantum experiment such as the double-slit experiment, it is the mind of the experimenter that determines its outcome.

One way that many people attempt to make sense of the counterintuitive aspects of quantum mechanics is to argue that there is some threshold between the quantum mechanical world of the very small and the human-sized classical world of the large. From this perspective, the strange things that we see in the quantum world simply don't apply to the classical world in which we live, which is why it seems we don't see these effects in everyday life (except, perhaps, in the form of weird phenomena). In practical terms, this means that if we need to make a calculation about some aspect of the physical world, we should use different tools depending on how big is the subject of our calculation. For example, if we are constructing microscopic electronic computer circuitry, then we need to use quantum mechanics but if we are calculating the path a football takes towards the goal after a free kick, then we would use classical physics (specifically Newton's laws of motion). As we see from brain-based theories of consciousness, neuroscientists generally regard the brain as lying above this threshold and hence treat it as a classical object where quantum mechanics is not relevant. This focus on neurons, brain structures, brain waves and neurochemistry obeying classical laws works well for many applications and has resulted in the huge technological advances in neurological medicine we've seen over the past few decades.

However, Stapp believes that attempting to explain consciousness with a classical model of the brain is doomed to failure. He argues that there aren't two supposedly separate 'quantum' and 'classical' worlds. Instead, the universe is entirely quantum mechanical. This includes, of course, our brains. If we look more closely at the brain, we will see that each tiny portion is in a quantum mechanical state. This has to be the case because the brain, like all physical matter, is ultimately composed of subatomic particles, which behave according to quantum theory. The reason why the brain is usually observed behaving in predictable classical terms is because the effects of quantum mechanics mostly cancel each other out. Stapp argues that because the brain is in fact made up of quantum states, this allows it to respond to the choices of a separate consciousness in the form of intentions. Hence consciousness performs the role of the observer in the Copenhagen interpretation, just as von Neumann suggested, and collapses wave functions in the brain.

The name von Neumann gave to this separate non-physical and mental aspect to the world that acts on the brain is the 'abstract ego'. It is our consciousness (our abstract ego) lying outside of physical space-time that is what causes the collapse of the quantum wave function from its multiple potentialities into a single brain-based physical actuality. It is also our consciousness that experiences this choice. At the stage of any conscious decision, there is a symmetrical correspondence between the physical state of the brain and its mental state meaning not only that changes in the mental (our consciousness) give rise to changes in the physical (our brain) but also that changes to the physical give rise to changes in the mental. All of this takes place through quantum effects. This is why we see the so-called neural correlates of conscious experiences (certain brain areas lighting up depending on our experiences) that are observed in the brain in imaging studies. It is also why alterations to the brain, for example because of a head injury or from ingesting certain substances, cause changes in our subjective experiences.

In a nutshell, then, Stapp's theory of consciousness builds on von Neumann's version of the Copenhagen interpretation to assert that a non-physical consciousness lying outside of space and time—the abstract ego—acts on the brain through making observations that collapse quantum brain states.

Features of Consciousness

How might these quantum approaches to consciousness address our three features of consciousness of rich personal experience, differing states of consciousness, and determining between conscious and non-conscious entities?

Rich Personal Experience

As we have seen, Hameroff and Penrose suggest that conscious experience is equivalent to an orchestrated collapse of the quantum wave function within neuronal microtubules. Whilst in a sense this is a brain-based process in that the brain is the place in which a non-local quantum collapse localises, their approach is fundamentally different to that of the brain-based theories of Chapter 2 because it is not based on neuronal interactions. Nevertheless, accepting the equivalence between entangled quantum collapse events and conscious experience still requires a similar leap of faith to that needed for the idea that neuronal brain processes produce consciousness, akin perhaps to believing there is "pixie dust in the synapses" as the philosopher Patricia Churchland graphically puts it.[182] However, this theory does have the virtue of suggesting that consciousness has a fundamentally distinct nature that cannot be expressed by the essentially computational and physical processes proposed by brain-based theories as equivalent to consciousness.

A different approach is offered by the quantum theory of Stapp and von Neumann. This theory proposes that consciousness manifests in the physical world through the abstract ego and so it starts with the assumption that consciousness by its very nature is rich experience. This means (arguably) that no further explanation for this feature of consciousness is possible or required. Consciousness is as it is in all its rich experiential complexity. Its personal and subjective nature comes from how each of us is essentially an individual abstract ego that is bound to our brain. (Although this rather begs the question of how a non-physical mental substance can affect physical matter, an issue known in philosophy as the interaction problem.)[183]

States of Consciousness

Hameroff and Penrose suggest that the distinction between conscious and non-conscious states can be made depending on whether there is an occurrence or otherwise of orchestrated quantum events.[184] Therefore, just as with the brain-based theories, the focus is on how to differentiate between the presence and absence of consciousness rather than considering a greater variety of states of consciousness such as the differences between waking, dreaming and meditation. How the theory may be applied for a more nuanced explanation regarding these and other states of consciousness is not clear.

The question of how Stapp and von Neumann's theory might account for different states of consciousness, even with regards to the basic distinction between being conscious and unconscious, does not seem to have been addressed.

Determining Between Conscious and Non-Conscious Entities

According to Hameroff and Penrose, a basic form of consciousness pervades the universe through ubiquitous wave function collapse. So, in some sense, everything is conscious. However, a higher level of consciousness such as we associate with humans and some animals—one that incorporates moments of understanding and meaning—can only arise when the quantum wave collapse can be orchestrated across a system like a brain. Whether or not an organism has the complexity to exhibit such orchestration is therefore the dividing line between an entity that is conscious and one that is not.

Hameroff and Penrose suggest that the capacity to exhibit this higher form of consciousness may have first emerged with the evolution of simple organisms possessing a sufficient number of neurons—perhaps three hundred or so—to allow for this orchestrated quantum collapse. Such creatures first appeared in the fossil record during the first part of the Cambrian period around 540 million years ago. Hameroff and Penrose speculate that this evolutionary innovation of orchestrated collapse and the basis for consciousness as we experience it may have been responsible for the relatively sudden proliferation of diverse lifeforms, an event known as the Cambrian explosion. It is possible therefore that 'tiny worms and urchins' living then may have exhibited some form of what they describe as "primitive consciousness, albeit exceedingly slow and simple but still with useful conscious moments".[185]

With respect to artificial intelligence (AI), Hameroff has said in an interview that silicon-based computers, no matter how complex, can never be conscious because they lack the capacity to host a collapse of the quantum wave function. This means that however impressively AI is able to simulate consciousness, none of these systems is actually conscious, because so far they have all been built using traditional computing architecture. By the same logic, this means that quantum computers, which are specifically designed to host quantum wave function collapse, may be able to express consciousness, especially if they are comprised of self-organising organic materials that can orchestrate quantum collapse through entanglement.[186]

Stapp and von Neumann's theory appears once again to be silent on the issue of determining between conscious and non-conscious entities.

Weird Phenomena

The suggestion that quantum mechanics is in some way directly related to consciousness, and perhaps also to weird phenomena, was anticipated by many of its eminent founders.[187] For example, Max Planck—the prime instigator in this field—wrote that quantum mechanics "teaches that there are realities existing apart from our sense-perceptions" and asked "is there, in the last analysis, some basically sound foothold for this belief in miracle, no matter how bizarre and illogical may be the outer forms it takes?"

Wolfgang Pauli was also open to the possibility of weird phenomena. He wrote in 1961 of what he described as "carefully conducted experiments" involving card-guessing and the like conducted by J.B. Rhine (discussed in Chapter 4), in which Pauli suggested that "there has been no evidence of mistakes". He concluded that "should the positive results in the still controversial realm of 'extrasensory perception' finally turn out to be true, this could lead to developments which cannot be envisioned today".

Niels Bohr, who, like Planck and Pauli, was also the recipient of a Nobel prize in physics for his contributions to developing quantum mechanics, wrote of the "synthesis of psychic phenomena connected with introspection... [which] in several respects [has] a formal similarity with that with which, to the great disquietude of many physicists and philosophers, we have met in atomic [quantum] physics". There are numerous other examples of physicists influential in the field of quantum mechanics, including Albert Einstein, who have expressed similar sentiments showing that they are at least open to the association of consciousness with quantum mechanics and the possibility of weird phenomena.

But what do the specific approaches to a quantum consciousness considered in this chapter have to say about the weird phenomena we discussed in Chapters 3–6, specifically, that consciousness can exist without a body, that it can transcend space and time and influence matter, and that we have free will?

Consciousness Beyond the Body

The two authors of the theory of quantum wave collapse in microtubules differ in their willingness to speculate on the implications of their approach for weird phenomena. Whilst Penrose appears pointedly to avoid such considerations, Hameroff is markedly less circumspect. Together with Deepak Chopra, Hameroff coins the term 'quantum soul' to describe the possibility of an individual personal consciousness being able to exist beyond the body and beyond physical death.[188] He points out that whilst the brain requires a relatively high amount of energy to conduct its cognitive processing, for example for electrical signalling between neurons, this does not apply to consciousness in the form of orchestrated quantum wave function collapse, which occurs at a much deeper level and needs much less energy. This means that consciousness might still be active even when its host brain is not.

Hameroff has spoken about what might happen when someone experiences clinical death, a medical term for cardiac arrest with the associated cessation of blood flow and breathing. He quotes studies that were performed on potential heart donors showing that there is a burst of brain activity during clinical death in the form of synchronised gamma waves (which are usually associated with peak levels of concentration) that last for between thirty and ninety seconds. Some dismiss the significance of this event as a 'last gasp' neuronal spasm or a mass die-off of neurons, but this does not seem to explain why there is what Hameroff and Chopra call a 'globally organised coherent synchrony'.[189] Hameroff suggests instead that this end-of-life brain activity could be process of the soul leaving the body. This state also coincides with the life-threatening situations in which many people report having had what are often profound and transformative near-death experiences. Hameroff says:

> What could happen, when the blood stops flowing and the oxygen is no longer available [is that] the quantum information in the microtubules... can dissipate to the universe at large... but remain entangled as a unitary entity and persist. And if a patient is resuscitated, it goes back inside [the brain].[190]

However, if the person is not resuscitated and their physical body irretrievably dies (known medically as biological or brain death) then Hameroff suggests that this 'unitary entity' of entangled quantum information—a person's 'quantum soul'—might be able to occupy microtubules in another organism in the process of forming and hence reincarnate in another body. Ultimately, he says that of course we just "don't know. [But] my position is that near-death, out-of-body, afterlife and reincarnation are not impossible... My own personal opinion is that they do occur".[191]

Stapp suggests that the survival of a person's consciousness is conceivable within his and von Neumann's interpretation of quantum mechanics. Stapp argues that an individual's consciousness, or the 'abstract ego', is distinct from the physical world and the physical body. He suggests that a person's mind interacts with the physical world via its body's host brain, with which it is 'specially tuned'. Given this distinction between mind and brain, it is therefore possible to conceive that the mind might survive the death of the brain and, as he puts it, "continue to have a stream of conscious experiences of some sort", such as is suggested by accounts of near-death experiences and might allow for out-of-body experiences. Regarding reincarnation, Stapp speculates in a similar way to Hameroff that, after the death of its former body, an individual consciousness might be able to attach itself to another brain through its 'probing actions'.[192]

Consciousness Transcending Space, Time and Matter

Hameroff is open to the existence of psychic phenomena and has said that "I think it happens... I've seen enough and heard enough and had my own experiences to think that [such phenomena do] occur".[193] His explanation for experiences such as telepathy is that they probably involve quantum entanglement to enable information transfer between physically separated and distant entities. He suggests that acceptance of extrasensory perception is in some ways a logical consequence of his and Penrose's theory that the orchestration of wave function collapse within the brain requires entanglement between different regions. He says:

> I think we need nonlocality, entanglement, in the brain, forget about outside the brain. Just this side of the brain to that side of the brain—you get this perfect gamma synchrony coherence. How do you do that? You can't do it by neural signals... So I think we need quantum entanglement in the brain to explain the issue of temporal binding... [and] all these electrophysiological effects.[194]

Hence it may not be such a leap to suggest that entanglement could be at work beyond the brain too.

In terms of seeing the future, Hameroff and Penrose suggest, albeit tentatively, that the results of presentiment experiments might be related to the fact that quantum entanglements, including those occurring within microtubules and beyond the brain, do not always follow conventional time ordering. As we saw in Chapter 4, presentiment is where some aspect of a person's physiology reacts in anticipation of a future event that cannot possibly be known by normal means. Certain effects related to a quantum-mediated consciousness might therefore propagate into the past as well as into the future. Hameroff and Penrose write that in this sense conscious experiences could be "*temporally non-local* with backward time effects". Hence consciousness may not "follow the normal rules of sequential time progression, so long as this does not lead to contradictions with external causality".[195]

Whilst precognition and presentiment challenge our commonsense notion of time always travelling from the past to the future, quantum physicists are beginning to find hard evidence that certain quantum events do indeed seem to involve time travel into the past. For example, a mind-boggling study conducted by researchers at the University of Toronto shows photons—particles of light—exiting a material before they have entered it.[196] If consciousness is indeed quantum in nature, then it is possible to envisage that it too may show similar features of time reversal and perhaps access information from the future.

For Stapp, the idea of phenomena that seem to transcend what we consider to be the normal rules of space, time and matter is only surprising if we consider the world to be classical in nature rather than entirely quantum as Stapp argues. In order to account for both spatial- and matter-related weird phenomena, Stapp proposes that, in certain psychological states, an individual's consciousness might be able to reach beyond their body and cause a collapse of the quantum wave function in the wider physical world. In this way, their abstract ego can access, or 'probe', distant information and so exhibit telepathy or can directly affect distant physical matter and so exhibit psychokinesis or energy healing. However, he concedes that for his theory fully to account for such psychic phenomena "many details would need to be developed in order to make a serious proposal".[197]

Free Will

The authors of both of the theories presented here argue that their approaches are wholly compatible with us having genuine free will. Hameroff suggests that consciousness is fundamental and it is therefore the mind that exerts free will over its physical environment (usually and most directly) through the brain. In his and Penrose's theory, this 'will' happens when a conscious thought, which is an instance of orchestrated wave function collapse, selects a particular state of orchestrated microtubules. This now-defined microtubule state causes particular neurons to fire and hence the brain activity we see associated with conscious will. As for arguments against free will that suggest that the brain operates along deterministic lines and essentially follows set algorithms, Hameroff points out that this has been shown to be false by Penrose's demonstration that the brain must be influenced by a non-computable factor, by quantum consciousness. Both of these lines of argument against free will are therefore refuted if we accept that consciousness comes first.

With regard to Libet's experiments that appear to show deterministic brain processes occurring prior to a conscious decision, Hameroff maintains that this can be explained through the nonlocality of time in the quantum world in a similar way as he and Penrose speculated that presentiment might occur. In other words, quantum effects allow for information to be sent backwards as well as forwards in time. This means that a conscious decision can spark processes that seem to happen prior to that decision being made through a retrocausal effect and hence provide an interpretation of Libet's findings that is compatible with free will.[198]

Stapp's theory also describes a situation where consciousness is fundamental and where the mind exerts its free will over the physical world. Stapp explains Libet's experiments by suggesting that there always exist many 'templates for action' in the world that represent all possible future event paths. When one of these actions is chosen by consciousness and causes a collapse of the quantum wave function, the potentials for paths not taken are destroyed from the historical record and leave only the one that was actually made. This is why Libet saw only the brain processes relating to the chosen decision of his participant. Brain processes relating to other decisions not taken are no longer present. Therefore the decision to act causes the body's readiness to act in the past.[199] Whilst this and Hameroff's retrocausal explanation are stubbornly counterintuitive, as we discussed at the beginning of this chapter, such strange implications are simply the nature of quantum mechanics.

Chapter 8

Consciousness is Hyperdimensional

In the last chapter, we discussed two theories based on quantum mechanics where consciousness might extend beyond the brain. In this chapter, we examine a perspective on consciousness that builds on the suggestion from some physicists that the universe may consist of more dimensions than the three spatial and one time dimension with which we are most familiar in our daily lives. These consciousness theorists suggest that the stuff of our mind may be related in some way to such extra, or hyper, dimensions.

Before we delve into these theories, let's look at what is meant by the term dimension and what it might mean for there to be higher dimensions. A dimension is used to express a movement or an extent in a particular direction. Our world appears to consist of three dimensions in space—left and right, backwards and forwards, and up and down. This is how we measure things and locate places. For example, the size (or extent) of a box is expressed in these three dimensions—its width, its depth and its height. In order to locate a particular place on a map, we provide a grid reference expressed in two dimensions. The height dimension is not usually required because places are normally found on the ground; however, knowing the height of somewhere is sometimes important, for example, if we're to find a hotel room on a particular floor.

In a similar way to space, time is also often thought of as a dimension, albeit one in which we seemingly move in only one direction—from the past to the future—unlike spatial dimensions in which (in principle at least) we have complete freedom to move. We sometimes need the dimension of time to locate an event; for example, if you and I are to meet in a café, we'll need the spatial location of that place but we also need to know what time we're due to meet.

In his theory of special relativity developed in the early twentieth century, Albert Einstein described our universe as four-dimensional, as consisting of three dimensions of space and one of time. This led to the notion of spacetime to show how these dimensions are intimately related. This theory established that space and time are not as static and objective as they appear to be. They can change in certain circumstances and therefore time and space are relative—time can actually speed up and distances can actually get shorter—hence the name of the theory. (Although we'd have to be travelling at extremely high speeds or dealing with sensitive technologies such GPS before such differences become noticeable.)[200]

Whilst both quantum mechanics and Einstein's theories of relativity revolutionised our understanding of the world and proved to be incredibly accurate models of reality, they haven't resolved all unanswered questions in physics. In order to make progress, scientists have since developed other theories, many of which have involved extra dimensions. Soon after relativity was established, physicists Theodor Kaluza and Oskar Klein developed a theory involving five dimensions that showed how the fundamental forces of gravity and electromagnetism might be related.[201] By the 1980s, a number of theories called string theories were able to offer accurate descriptions of elementary particles, but only through invoking a universe consisting of ten dimensions—nine of space and one of time. These theories suggested that reality at its deepest level is formed from miniscule strings that manifest as fundamental particles and forces when vibrating at different frequencies.[202]

In the 1990s, physicists combined a number of these string theories into one overarching theory that they called M-theory. If M-theory is correct, then we live in a universe that consists of eleven dimensions. (What the 'M' in its name is said to stand for is somewhat contested, but it has been suggested that it may refer to 'membrane', which is what the tiny tubular strings would look like if they were stretched out but it could also mean 'mother'—as in the Mother theory—or even 'mystery'—as in 'who knows?!')[203] M-theory is not yet a comprehensive theory but some physicists believe it to be the closest we have come so far in defining a single set of mathematical equations to describe all of physical reality, in other words a so-called Theory of Everything.

Whilst eleven dimensions are necessary for the mathematics of M-theory to describe how fundamental forces and particles work, our everyday world does not appear to be hyperdimensional. Where are these extra dimensions and why is it that we seem not to experience them? The reason usually suggested is that the spatial dimensions in M-theory beyond the three with which we are most familiar are wrapped-up so small that they can't be discerned by our normal human senses nor can they be probed by scientific instruments, which can only operate in four-dimensional spacetime.[204]

Consciousness is Found in Hyperspace

The physicist Bernard Carr has another take. Perhaps, he suggests, these dimensions are not inaccessible to us and we do in fact routinely access levels of hyperspace. He argues that our consciousness might be found in these extra higher dimensions and that our minds explore these dimensions when we engage in any type of mental activity.[205,206,207,208]

Carr's argument stems from the observation that many mental activities appear to take place in environments that could be thought of as equivalent to physical spaces. For example, we can visualise in our mind's eye a mental space with objects that have positions relative to each other, such as picturing a bedroom with walls and a ceiling that enclose a mental space into which we might imagine a bed, a lamp and so on, all of which take up different locations in the space. Dreams too seem to take place in mental spaces in which we move from one place to another and interact with other characters who may be close or distant, large or small, albeit that these spaces seem more elastic and malleable than three-dimensional physical space. Nevertheless, such relative characteristics of extent are all properties of spaces. More exotic states of consciousness, such as the out-of-body experience, also seem to occur in a space-like environment, one that is reported to have close similarities with our waking world in which those who experience such states suggest that they can move around, view objects from different perspectives and travel from one mental location to another.[209] Many reports of near-death experiences also have space-like properties, as seen, for example, in accounts of travelling through a tunnel towards a light.[210]

Carr suggests that these higher dimensions might be informational consciousness spaces, with each ascending dimension corresponding to an increasing dimension of information. If this model is correct, then there needs to be some way for information to be shared between different dimensions. For example, our three-dimensional brains and mouths are clearly necessary for telling others about a dream or an out-of-body experience that we might have had in a higher dimension of reality. Hence there is a need for information to pass from one dimension to another. To accommodate this, Carr has proposed what he calls 'transcendental field theory', which is so-called because the interactions between dimensions might happen via fields and because these fields transcend – go beyond – our regular three spatial dimensions. Carr suggests that such information interactions between dimensions might be somehow geometric in form. In other words, this hypothesised field is also space-like.

Carr proposes that a hierarchy of physical dimensions might correspond to a hierarchy of mental spaces. In Carr's schema (which he admits is somewhat tentative), he suggests that the mental spaces that are uncontroversial to mainstream science – such as those involved in our perceptions of physical space, our memories and our dreams – might occupy a relatively low dimensional space, perhaps with five spatial dimensions. Certain aspects of phenomena that seem to challenge our normal notions of spatial or temporal relationships such as telepathy or precognition might be explained in terms of still higher dimensions of space or time. These extra dimensions could allow for apparent violations of rules that seem otherwise to be sacrosanct to our familiar physical world. How higher dimensions could explain such weird information transfer might perhaps best be illuminated through the story of Flatland, a fictional two-dimensional world imagined in a Victorian novella of that name by Edwin A. Abbot.[211] By considering a world of only two dimensions from our perspective of three spatial dimensions, it becomes possible for us to conceive of some of the characteristics that might transpire in a hyperdimensional world.

Abbot introduces us to a character called A. Square (who is, unsurprisingly, a square) who lives in a world in which everything exists on a flat plane. People in Flatland are two-dimensional shapes and so can only move around the plane in combinations of right-and-left and backwards-and-forwards movements. There is no up-and-down in Flatland. The walls of buildings are like lines on a sheet of paper. The only way for A. Square to enter or leave his two-dimensional house is if there is a gap in one of these lines, such as for his doorway. However, there is a way to get into the building when the doors are locked and without breaking a wall that is inconceivable (at the start of the story at least) to A. Square and this is by invoking a third dimension of height. In the novella, A. Square is visited by a character who exists in three dimensions. This character is called A.

Sphere (yes, he's a sphere). Because Square can perceive only two dimensions, Sphere appears to him as a circle. And, depending on Sphere's up and down movements interacting with Flatland, this circle seems to grow and shrink. From the viewpoint of Square, Sphere can even apparently disappear entirely and magically reappear in another place by raising himself above the plane of Flatland, moving, and then descending somewhere else. So, by using three dimensions, Sphere can easily move into and out of a building in Flatland that appears to be securely locked. This feat is impossible using only two dimensions but becomes straightforward when using three.

Another implication is that things that seem to be spatially separated in two dimensions may in fact be part of a single three-dimensional object. For example, consider the legs of a three-dimensional chair that someone might be holding above Flatland. As they lower the chair and it intersects the plane of Flatland, the legs appear as a group of four apparently separate objects that nevertheless seem to be mysteriously related as the person moves it around. If the chair is lowered until the flat seat intersects with Flatland, then the four objects suddenly become a single much larger object. If we lived in only two dimensions, such weird phenomena would be difficult to account for. However, they make perfect sense from a three-dimensional perspective.

By extending such a model to our world, if mental activities take place in higher dimensions than three, this would allow for information to transfer in ways that otherwise seem impossible. Such higher-dimensional information transfer could potentially account for otherwise mysterious processes like telepathy. A multidimensional model in which consciousness is not located in (three-dimensional) spacetime but is capable of exchanging information with spacetime objects might also explain certain phenomena associated with survival after bodily death, such as reincarnation. Higher-dimensional spaces might be shared and communal and allow a disembodied consciousness to live independently of the physical body with which it was formerly associated. Higher dimensions could also explain otherwise inexplicable behaviour exhibited by ghosts that seem able to defy three-dimensional physical rules such as suddenly appearing or disappearing, or being able to walk through walls. Mystical experiences, which are often reported to involve feelings of unity or oneness, might be tapping into what Carr calls a Universal Structure that encompasses every dimension.

As well as spatial dimensions and transcendental fields for the transfer of information, time is also integral to Carr's model. He suggests that physical time, which can be measured using a physical clock, might be distinct from the mental time that we subjectively experience as individuals. This would accord with how we can have an experience of time that doesn't always match the regularity of a timepiece with an hour passing quickly when we are enjoying ourselves but feeling like an eternity when waiting for some important event. As such, this mental time is an important component of how we feel as individuals. Carr proposes that for each 'plane of actuality', such as the dimensions in which we dream or experience out-of-body experiences, there is a separate expression, or 'projection', of time. He writes, "it is as though consciousness perceives the world through a number of windows, each with its own clock".[212]

In particular, Carr suggests that our subjective experience of the present is not confined to a single momentary instant but actually extends over a period with a distinct duration. In this longer period, separate sequential occurrences such as rapidly flashing lights or musical notes are actually perceived as happening at the same time. This has been termed the 'specious present',[213] specious in the sense that conceiving of the present as a single moment is misleading. The specious present might be thought of as a granularity or a pixellation of time—an indivisible unit of experienced, subjective time. Some have suggested that the specious present in normal states of consciousness might be between around 0.03 seconds (which relates to the speed of oscillations in our brain) to around 0.1 seconds (which is the minimum timeframe our senses need to observe an event). Carr suggests that the specious present may change substantially in unusual circumstances such as during paranormal or spiritual experiences. He proposes a hierarchy of increasingly long specious presents corresponding with our consciousness occupying higher spatial dimensions. This may explain why those involved in a traumatic event, such as a car crash, report things happening in slow motion, why some of those who have had near-death experiences report re-experiencing their whole lives in an instant and why mystical states are sometimes associated with a feeling of timelessness.

A Trinity of Space, Time and Consciousness

Another theory of consciousness based on the idea of higher dimensions has been proposed by physicist, cosmologist and mathematician

Edward Close and neuropsychiatrist Vernon Neppe. Close and Neppe's theory is a highly ambitious proposal to explain the entire universe with all its physical and mental aspects through precise mathematical principles. They describe their theory as the Triadic Dimensional Vortical Paradigm. The model is triadic because of the suggestion that the universe is comprised of three basic dimensional elements or substrates—namely, space, time and consciousness.

Their schema integrates infinite consciousness as a fundamental component of reality without which, they argue, theories in physics are destined always to be incomplete and insufficient to explain the universe. Instead, they propose that, alongside the three dimensions of space and one of time that we normally experience, there are two additional dimensions of time, making three temporal dimensions in all, together with at least three dimensions of consciousness. These dimensions are bridged through trans-dimensional dynamic vortices to comprise a holistic physical and mental reality. They claim that theirs is a true Theory of Everything because it goes beyond Theories of Everything in physics—such as M-theory—which are limited to attempting to explain the physical world in solely physical terms and ignore consciousness altogether.[214,215]

The idea of space having three dimensions is perhaps the most straightforward component of Close and Neppe's theory. That there really are only the three spatial dimensions with which we are familiar accords with commonsense and how we daily experience moving around and interacting in space. Theories of physics that invoke higher dimensions of space in order to address particular mathematical enigmas are dismissed by Close and Neppe as 'simple conceptual tricks'.[216] They suggest that their theory does away with the need for such intellectual gymnastics and hence is a much more parsimonious approach.

The idea of time as also having three dimensions is at first glance somewhat less intuitive. However, Close and Neppe's mathematical model—expressed in algebraic and geometric terms—suggests that there are three temporal dimensions and so the challenge comes in explaining what this might actually mean. Our lives can be seen as comprising an ongoing series of singular present moments. From this perspective, we experience each moment as point-like and therefore having zero dimensions. Some spiritual traditions have emphasised this idea as a route to more enlightened states through meditation techniques that focus on placing one's awareness on the present, such as in the widespread practice of mindfulness that originated in ancient

Hindu and Buddhist teachings and has been promoted in the West by individuals such as Jon Kabat-Zinn[217] and Ekhart Tolle in his best-selling book *The Power of Now*.[218] However, our having memories of the past alongside potential knowledge of future events through weird phenomena such as precognition and presentiment might suggest that our experience of time is more like a one-dimensional line than point-like because it extends both forwards and backwards. As to the second dimension of time, Close and Neppe propose that this extends at right-angles from the first dimension of time at each present point-like moment and is traversed through choices made in our lives. Hence this dimension could manifest as alternative time-lines representing a multitude of possible worlds that we experience depending on decisions we make. The third time dimension is proposed to be that which couples the first two dimensions of time to the consciousness substrate through us interacting with others and thereby creating a three-dimensional volume to time.

The final, and most consequential, dimensional component of Close and Neppe's theory is that of consciousness. As well as three spatial and three temporal dimensions, they posit that reality includes a substrate of consciousness, an underlying 'metaconsciousness' that is potentially infinite. However, our individual experience of consciousness is usually not infinite and is instead expressed in more limited dimensions. For example, they suggest that, at the lower end of these dimensions, metaconsciousness might take the form of information, particularly in our idea of meaning. At higher dimensions, metaconsciousness may be expressed, as they put it, "with qualities such as honesty, love, hate, modesty, satisfaction, wisdom, understanding".[219] Metaconsciousness may also find expression across groups of individuals, such as families, nations and ethnic groups. Whilst Close and Neppe's mathematical model points to a minimum of three dimensions of consciousness, they suggest there might potentially be an infinite and varying number of consciousness dimensions, depending on an individual's—or a group's—experiences and circumstances. Fluctuations in the number of dimensions could reflect different states of consciousness and so particular dimensions of consciousness may be lost when this state changes, for example, upon waking from a dream or concluding a meditation.

The substance of consciousness is expressed in Close and Neppe's mathematical model in the form of what they call 'gimmel'. (Gimmel is the third letter of the Hebrew alphabet and Jewish rabbinic sources describe the meaning of the word as a bridge between the physical world and God.)[220] Close and Neppe use the term gimmel to retain some neutrality on the question of whether their model points to consciousness itself or instead some 'vehicle' or 'carrier' for consciousness. Gimmel is without mass and energy but can apparently explain why physical particles retain their structure and order. They suggest that, "without gimmel, the spinning (vortical) atoms would be unstable and asymmetrical about their axes and would, in effect, fly apart: Our world and the physical universe could not exist".[221]

So we have a model of reality that comprises three spatial, three temporal and at least three consciousness dimensions. However, whilst this model describes the context for reality, it does not account for its content. This is provided for in Close and Neppe's model through vortices, dynamic three-dimensional entities, which they suggest are the "predominant structural form of nature".[222] Examples of physical vortices include whirlpools in water or tornadoes in weather systems. However, Close and Neppe propose that vortices are ubiquitous throughout reality and facilitate the transmission of information across and between dimensions of space, time and consciousness. Vortices can also be entirely constituted from consciousness with no spatial or temporal components. The interaction of vortices with each other provides for the integration of dimensions through what Close and Neppe call 'tethering', ultimately allowing for the holistic nature of reality.

Close and Neppe claim that their theory can explain the presence of life. There is an established principle in physics known as the second law of thermodynamics. This states that everything will tend towards disorder, known technically as entropy. However, the existence of life and evolution seems to contradict this law through the presence of highly ordered and complex living systems. This increase in order is usually explained in conventional physics as being balanced by a greater amount of disorder that is created as a consequence of life's processes, thus retaining the basic principles of the second law when viewed as a whole. However, Close and Neppe point out that the same conventional scientific paradigm has no explanation for the origin of life. This essentially reductionist paradigm can explain many aspects of the processes of life, but not its creation. In Close and Neppe's model, life is able to utilise consciousness and create order. Whilst consciousness is not synonymous with life, living sentient beings are nevertheless seemingly required to experience conscious awareness in space and time.

Close and Neppe suggest that their triadic dimensional theory of the universe contains an infinite potential for life through a close relationship with the presence of consciousness. This is not simply idle speculation without theoretical basis. In their paper published in an international physics journal,[223] they describe how in their model the elements that are necessary for life, such as carbon, oxygen and hydrogen, have more gimmel than other elements and that water—the single most vital substance for life—has more gimmel associated with it than any other compound. As well as the presence of gimmel in life-essential elements, Close and Neppe suggest that their mathematical model can also account well for other currently unresolved questions in science such as models of dark matter and dark energy, as well as data emanating from the Large Hadron Collider at CERN concerning the nature of fundamental particles.

Features of Consciousness

How might these hyperdimensional approaches to consciousness address our three features of consciousness of rich personal experience, differing states of consciousness, and determining between conscious and non-conscious entities?

Rich Personal Experience

Carr's theory is focused on the operations of consciousness in higher dimensions of space. Hence, why and how consciousness might express itself as rich personal experience is not addressed.

The triadic dimensional theory of Close and Neppe proposes that a substrate of multidimensional consciousness is fundamental to reality along with the substrates of space and time. From this perspective, then, the nature of consciousness as rich phenomenal experience requires no explanation. Although this is not explicit in their theory, it's possible to surmise that our feeling that experiences are subjective and personal to us happens because our usual experience of consciousness is through lower dimensions of consciousness that appear separated. If we access higher dimensions through altered states, then these levels of consciousness are shared with progressively wider networks of people (family, sports teams, nations), which are less uniquely personal.

States of Consciousness

It's possible that Carr's theory could explain different states of consciousness through the mind engaging in travel to different locations within higher dimensions of space that encompass different qualities associated with these states. Similarly, Close and Neppe's theory might explain different states of consciousness through travel across multiple consciousness resulting in different experiential qualities. This is how the theories might account for the unique qualities that are exhibited whilst in each state. However, because little attention appears to have been paid to this question by the authors of the theories themselves, this must for now remain somewhat speculative.

Determining Between Conscious and Non-Conscious Entities

These hyperdimensional theories appear not to address explicitly the issue of how we might distinguish between conscious entities and non-conscious entities. There also seem to be no obvious implications from the theories that could help us to do so meaning that this question remains unanswered.

Weird Phenomena

The potential reality of dimensions beyond the three of space and one of time in established theories of physics opens the possibility for novel explanations of the weird phenomena we examined in Chapters 3–6. In particular, as we saw, weird phenomena are an intrinsic part of Bernard Carr's theory of consciousness. Indeed, Carr's belief in the likelihood of these as genuine is a large part of what motivated him to develop his hyperdimensional theory.[224] Explanations for weird phenomena also seem to fit well with Close and Neppe's perspectives on reality. Let's examine how these theories account for these weird phenomena.

Consciousness Beyond the Body

Carr suggests that higher-dimensional physical spaces could accommodate an individual's consciousness even when it is no longer attached to a physical body. Some of these dimensions might constitute what he calls an after-death 'survival space', the travel to which through a dark tunnel is sometimes reported by those who have had near-death experiences. This survival space could be where souls are located between physical incarnations, perhaps equivalent to the bardo or intermediate state that is described in Buddhist traditions and which is sometimes recalled by people who say they have experienced an existence between lives in reincarnation cases.[225] It may be that this space is what is referred to as heaven in the Christian tradition. Carr believes that survival after death may even be a logical implication of his theory because, in his schema, consciousness occupies mental spaces that go beyond the confines of the physical body and operates in higher spatial dimensions, each of which has its own related subjective perception of time—its own specious present. As Carr puts it, "the mind is 'larger' than the body and so may persist outside it and endure longer than it".[226]

Close and Neppe's theory also allows for consciousness beyond the body. They argue that the substrate of consciousness contains an infinite potential for life that might be similar in some respects to the Christian idea of a soul or, perhaps more accurately, the Indian concept of *atma* found in philosophies such as Hinduism and Buddhism. With its meaning as 'breath' or 'essence', *atma* has a less individualised connotation than Western religious traditions' concepts of soul. Close and Neppe speculate that this potential-for-life found in dimensions of consciousness might be able to interact with space and time dimensions through the form of a physical body and so be able to direct its actions during its lifetime. After the death of its body, the soul/*atma* would cease to have any ongoing connection with spatial and time dimensions and therefore normally persist only within dimensions of consciousness. Apparent communication with people who have died, whether spontaneous or through a medium, are only brief connections between the soul/*atma* and our four-dimensional perceptual world. The potentially infinite nature of the consciousness substrate may also, Close and Neppe suggest, explain why our understandings of metaconsciousness are so limited when we try to express them in finite dimensions of space and time.

Consciousness Transcending Space, Time and Matter

Weird phenomena that challenge our commonsense notions of space and time become simply a matter of perspective if consciousness is to be found in higher dimensional space as in Carr's theory. In four-dimensional spacetime our brains are entirely isolated within our skulls, physical places are separate and we have a sequential experience of time. However, in higher dimensions information that appears discrete in spacetime might be shared in a similar way to how the legs of a chair appear to be four entirely separate objects in two-dimensional Flatland but joined together from the point of view of an extra (third) dimension. This means that telepathy becomes feasible, perhaps even expected, when we mentally occupy shared dimensions beyond four-dimensional spacetime.

As for knowledge of the future, perhaps this is made possible through extensions of the specious present. Events that appear to be ordered and distinct in our usual experience of time become seemingly simultaneous if they occur within the same granular unit of time. Hence the length of a normal specious present for humans may be enough to accommodate a relatively short presentiment effect such as that seen in experiments described in Chapter 4 that show that physiological systems over which we have no conscious control (such as our pupils dilating) can respond to events that have not yet occurred. And precognition over longer periods of time might be possible through significantly longer durations of the specious present, for example when in a dream-like state or some other altered state of consciousness.

Close and Neppe also regard phenomena that appears weird from our commonly-experienced four-dimensional spacetime as entirely normal and expected when understood from a higher dimensional perspective. They suggest that phenomena such as telepathy are instances of information transfer between the consciousness of humans, animals or other organisms facilitated by the movement of vortices within dimensions of consciousness. Presentiment and precognition might be better explained with second and third time dimensions allowing greater potential freedom of temporal movement rather than assuming we inhabit a universe that allows for only one-way travel along a single dimension of time as conventional physics would have it. Higher dimensions of time might also explain the seemingly bewildering reversal of 'normal' time ordering that we see in some quantum experiments, such as that mentioned in the previous chapter where photons were observed leaving a material before they have even entered it.[227] As regards the influence of mind over matter through psychokinesis and energy healing, again this could be accounted for by the movement of vortices, this time between an individual expression of consciousness and physical matter.

Close and Neppe speculate that this weird information transfer between dimensions may involve special particles—which they term 'psitrons' (for psychic information transfer) and 'kinetrons' (for mind-over-matter)—in a similar way that fundamental physical forces like gravity and subatomic forces involve particles such as leptons and hadrons.

Close and Neppe point out that weird phenomena are relatively rare occurrences, experienced and reported only infrequently. This may be why most parapsychological experiments show relatively small effect sizes. In order to explain why weird phenomena are not more common, Close and Neppe point to the filter theory of the brain first proposed at the beginning of the twentieth century by prominent intellectuals including William James and Henri Bergson and later popularised by the author Aldous Huxley.[228] This idea suggests that the brain acts as a filter for a more extensive consciousness, such as the metaconsciousness of Close and Neppe's theory. The brain works to severely constrain the information that is available to an individual's conscious awareness. This is necessary because otherwise the brain would be overwhelmed with information and this would limit our capacity to act effectively in the world. The brain-working-as-filter enables us to prioritise what is immediately biologically useful for our survival. This is needed not only for 'normal' sensory information—we are consciously aware of only a tiny proportion of what our senses are picking up—but it also offers us an explanation for why we are only seldom privy to psychic information. This may also explain why it is that when we are asleep, in meditation or under the influence of hallucinogens we are more receptive than when we're awake to phenomena such as telepathy or precognition ("communication between the infinite and the finite", as Close and Neppe put it)[229] because these states are all associated with a reduction in conscious thinking.

Close and Neppe also suggest that individuals who have close bonds, for example people from the same family, are more receptive to sharing psychic information because they have closely intersecting vortices. As well as emotional bonds, they suggest that cognitive bonds may be important too, relating how they have noticed remarkable synchronicities between them where each has arrived at the same idea or concept individually at the exact same time despite not being in contact through conventional means:

> We would often call each other at opportune times and find that we had been thinking along the same lines. We would wake up in the middle of the night with epiphanies that were nearly identical, or that would fit together in complementary ways. The frequency of such synchronous experiences led us to suspect a hyper-dimensional connection of some sort. As we developed our new paradigm, the

possibility and reality of such a link became more and more apparent. When consciousness is seen to operate in more than four dimensions, synchronicity becomes the rule rather than the exception.[230]

Free Will

How might these hyperdimensional theories accommodate free will? First, as we discussed in Chapter 6, it is important to remember that our currently-accepted theories of physics are thought of by many as being incompatible with us having genuine free will. In particular, Einstein's four-dimensional spacetime is often presented as a 'block universe' with our past, present and future all predetermined, coexisting and unalterable.[231]

However, there is an issue with this model. Carr points out that there is no concept of the 'now'—our perception of the present moment—in the block universe. As Carr notes, mental time as we experience it must therefore somehow be different from the physical time in Einstein's theory. It could be assumed that our consciousness encounters events in the block universe as it makes its way through spacetime but nothing in Einstein's theory of relativity describes or even hints at such a process. Carr concludes from this that there must be at least one extra dimension to those offered by Einstein that describes how consciousness moves through physical time in the four-dimensional block universe. This fifth dimension could take the form of the many potential futures encapsulated in the quantum mechanical wave function (discussed in Chapter 7), one of which is selected by our consciousness to become what actually happens. Or perhaps all of these possible futures become real as in the many-worlds interpretation of quantum mechanics (we'll look at this interpretation in a bit more depth in Chapter 11), in which case, perhaps our consciousness selects just one of these many worlds to experience. Either way, Carr's hyperdimensional model of physics is compatible with free will.

In Close and Neppe's theory, time has three dimensions as derived from their mathematical formulation. This model fits well with the idea of us having free will. As we discussed earlier in this chapter, the first time dimension can be thought of as a timeline along which can be found past, present and future. The second time dimension extends from each point of the 'now' giving us a time structure that resembles a flat plane. This is experienced through us exercising our free choice at each present moment so that we inhabit a two-dimensional time structure, albeit without the freedom to explore as we can in two-dimensional space. The third time dimension is encountered when we interact with others who are also exercising their free will. This model therefore allows free will as it describes how we may move through temporal avenues that would otherwise be impossible in a world with just one dimension of time, or even zero dimensions in the form of the point-like present moment.

Chapter 9

Consciousness is Field-Like

In our tour of theories that suggest consciousness reaches beyond the brain, we have so far discussed those with explanations that involve quantum mechanics and those that involve higher dimensions. In this chapter, we'll look at theories that suggest that consciousness may be associated with fields. But what exactly is a field in a scientific context? A field is an area that's influenced by whatever is producing the field even if that something is far away. For example, gravity exerts a field. The gravitational field of the Earth is what keeps objects on the ground. We can throw something in the air but, as the saying goes, what goes up must come down, unless we throw it with enough force that it can escape the pull of the Earth's gravitational field that keeps it here, for example by blasting it in a rocket. The influence of a gravitational field can extend for extremely long distances. For example, the most distant objects orbiting the Sun are around 15 billion kilometres away.

Consciousness Uses Electromagnetism and Biofields

The idea that consciousness might be field-like can be compatible with the premise of theories that suggest the brain is the source of consciousness such as those discussed in Chapter 2. There are a number of theorists who have proposed that consciousness might be found in the electromagnetic fields that are generated by the brain and the body. This suggestion is gaining in popularity in the scientific community because field theories of consciousness seem to overcome some of the objections raised from other theories of brain-based consciousness.[232] For example, they might be able to explain why consciousness feels unified, because fields are by their nature unified.

This unity is difficult to account for when consciousness is said to emerge from discrete neurons and synapses. However, as with other brain-based theories, most advocates of field theories have not addressed how their proposals might accommodate weird phenomena.

One exception is integrative medical doctor and neuropsychiatrist Diane Hennacy (also known as Diane Hennacy Powell). Hennacy proposes that the brain's primary role is to interface with and navigate a wider consciousness, in a similar manner to how computers interact with information in the cloud. She suggests that examining how the brain works could provide clues as to how it might use electromagnetism to interface with a holographic non-local consciousness that could perhaps usher in a revolution for neuroscience—'Neuroscience 2.0', as she puts it.[233,234]

What leads Hennacy to suggest that electromagnetism might play a major role in processing consciousness is that electromagnetic fields are involved in a number of aspects of our conscious awareness. For example, in terms of our senses, our eyes receive visible electromagnetic light waves and our sense of touch is essentially an awareness of electromagnetic forces in the atoms that make up the cells in our skin being repelled by the electromagnetism in the atoms of whatever we're in contact with. The activity of the brain—the firing of neurons and synapses, for example—creates electromagnetic waves that can be detected outside of the skull using an electroencephalogram (EEG). Different frequencies of these brain waves are associated with different types of brain activity. For example, if you're deep in concentration your brain will produce electromagnetic frequencies greater than 35 Hz, known as gamma waves. If you're in deep sleep, your brain will produce frequencies less than 4 Hz, known as delta waves.

When we think of brain cells, we might first think of neurons. Neurons constitute the grey matter of the brain and this type of cell has been the focus for much of the history of neuroscience, hence the name for this field of research. However, 90% of cells in the brain are glial cells, not neurons. Whilst neurons communicate with each other using chemicals, the primary means used by glial cells for communication is electrical, which is much faster than chemical-based processing. Some have suggested that glial cells might be associated with our higher thought processes such as imagination and creativity, potentially making them more important with regards to consciousness than neurons.[235]

Unlike most of the scientists who have proposed electromagnetism as potentially underlying consciousness, Hennacy argues that these fields may be important within a model of consciousness that is not confined to the brain. In particular, she proposes a role for Extremely Low Frequency (ELF) waves. ELF waves have frequencies between 3 and 30 Hz with corresponding wavelengths of 100,000 to 10,000 km. This means that ELF waves can be transmitted over an extremely wide area and pass through most physical structures, even those structures specifically designed to shield electromagnetism such as Farraday cages. Hennacy notes that the brain waves most associated with dreaming and deep relaxation are theta waves, which have frequencies of 6 to 10 Hz. In other words, they are ELF waves.

There is some evidence that animals might be able to detect ELF waves, for example to support long-distance migratory behaviour through making use of the Earth's magnetic field, perhaps with a compass-like role for the magnetite (the most magnetic of all minerals) that is found in birds' beaks. Humans also have magnetite in the nose bridge and near the temples so it is possible that we too may use this as a kind of antenna for ELF waves to help with our sense of direction. Apart from these concentrations in the skull, most of the magnetite found in the human body is in the tissues that surround the brain. It's possible that this magnetite might act as a shield against external electromagnetic radiation.[236] Perhaps, as we'll discuss in more depth later in this chapter, this magnetite acts to prevent us becoming overwhelmed by information from the consciousness of others or from universal consciousness, which would fit with the brain-as-filter idea that we discussed in the context of Close and Neppe's theory in the last chapter.

There are a number of other aspects to the brain and body that Hennacy suggests might be important in relation to electromagnetic effects and therefore to consciousness. These include neuromelanin, the pineal gland, microtubules and mitochondria as a source of biophotons. Let's look at each of these in turn.

Neuromelanin is found only in neurons, and the colouration it gives to cells is what gives rise to the term grey matter. (The glial cells we discussed earlier don't have melanin and so are referred to as white matter.) Melanin is sensitive to electromagnetic fields. For example, it absorbs a wide range of wavelengths of light, which might mean we are able subconsciously to be aware of a wider spectrum of the electromagnetic spectrum than just 'visible' light. Melanin can also respond to sound waves. Hennacy suggests that one implication of these associations with electromagnetism is that melanin might be able to act as what she calls a 'biological semiconductor' enabling our bodies to respond to light, sound and electrical stimuli that are below our conscious awareness. Hennacy speculates that the presence of neuromelanin in neurons may therefore offer the potential to develop technology to affect states of consciousness through the application of light and sound waves "like buttons on a remote control", somewhat analogous to the role of semiconductors in electronic components.[237]

Hennacy suggests that the pineal gland might be involved in information processing due to its central location within the brain so that any signals it may emit can speedily reach all regions of the brain. The pineal gland is encased in greater amounts of electrolyte fluid, an electromagnetically-conductive material, than any other part of the brain. The pineal gland is highly sensitive to light and other electromagnetic fields and produces melatonin, which is important in sleep regulation. The philosopher Descartes suggested that the pineal gland was where the soul might be found and from which thoughts emerge.[238] It has also been suggested by some to be involved in psychic phenomena,[239] providing further evidence that it may have a role in interactions with a wider, shared, consciousness.

In Chapter 7, we looked at Hameroff and Penrose's theory that consciousness equates to an orchestrated collapse of the quantum wave function in microtubules. Hennacy agrees that microtubules may be related to consciousness but, instead of being host to quantum processes as Hameroff and Penrose suggest, she proposes that microtubules may act as electromagnetic superconductors and so help the transmission of electromagnetic fields throughout the brain. The key, she argues, is in their tubular shape, which is three hundred times more conductive of electricity than the protein material it's made of would be if it had no particular arrangement.

Mitochondria are found close to the brain's microtubules. Whilst their role in the brain has long been established as an energy source for synapses, it has been suggested that these structures may also be a source for biophotons. Biophotons—biological particles of light—are emitted by every living organism, meaning that life quite literally glows, although we can't see these biophotons because they are much too few in number for our eyes to detect. Whilst most biophotons are used for various physiological processes, Hennacy cites research by the physicist Claude Swanson who suggests that the emission of biophotons may create holographic torsion fields within and beyond the body, and it is these that perhaps equate to a person's 'aura' or 'subtle body', suggested by some to surround us and offer clues to our state of being. The word torsion is related to the twisting effect by which the field is produced. And it is said to be holographic in the sense that it may relate to higher dimensions in the same way as a two-dimensional structure can play host to a three-dimensional holographic laser-generated image.[240] Hennacy suggests that these holographic torsion fields could be the means by which our bodies can access non-local information from an expansive consciousness beyond the brain.

Consciousness is Found in Morphic Fields

Another theory that suggests consciousness may be found in fields has been proposed by the biologist Rupert Sheldrake. Whilst Diane Hennacy's theory relates to electromagnetism, which is a type of field familiar to us in our daily lives, Sheldrake's involves fields the existence of which is more speculative, fields that he calls morphic fields.

In common with Hennacy and other theories of non-local consciousness, Sheldrake maintains that the brain is not the source of consciousness. Instead, he suggests that there are non-local fields of consciousness that extend beyond the body with which we resonate. Just as a mobile phone does not of course have within it the person we're speaking with but instead tunes in to electromagnetic signals broadcast from a mobile phone mast, so he believes that our brains are able to tune into and access morphic fields of consciousness that are beyond the body. Sheldrake is not the first to suggest this notion of the brain as similar to a radio receiver. For example, in the 1930s, the popular author and politician Upton Sinclair published a book called *Mental Radio* (which included a short preface from Albert Einstein), drawing on a similar metaphor to explain telepathy where he compared our brains to radios that pick up broadcasts (that is, thoughts) from afar.[241]

Sheldrake first proposed his concept of morphic fields to provide an explanation of how lifeforms are able to form physical structures with precise and distinct shapes and compositions during their development. This would explain, for example, how cells are able to organise into the various complex structures of a heart or an arm or a nose or skin using a kind of master template embodied in the morphic field. Whilst most people assume that this process of organ and whole organism formation can be explained using genetics, Sheldrake argues that, because each cell in the body shares the same genetic code, it is hard to envisage how the same genes in cells with drastically different functions could determine such behaviour.

Instead, Sheldrake hypothesises that there are morphic fields that hold a blueprint for the physical shape and structure of organisms and their component parts. Cells use these fields to guide the relative positioning and particular composition of every part of the body. Each specific entity within the body—whether a tissue, organ, limb or the whole organism itself—has its own morphic field associated with its formation that has evolved as previous organisms of the same species have both contributed to and accessed these fields, strengthening and refining them. This is similar to how water flowing down a hill will eventually erode a channel. This channel becomes more established over time as the water is both guided by the channel but also makes it more pronounced through further erosion so making it more likely that future flows of water will follow the same channel.

Sheldrake's proposal is that morphic fields apply well beyond biological development to encompass many aspects of the universe, including accounting for the laws of nature, although he suggests that this "law metaphor is embarrassingly anthropomorphic".[242] Instead, he proposes that the so-called laws of nature might be better thought of as entrenched habits rather than universal principles. A morphic field associated with a new physical behaviour will be, according to Sheldrake's theory, initially weak before it gradually settles down and becomes stronger. One example he cites that is subject to empirical observation is novel crystalline compounds that are made in the lab.

Sheldrake reasons that these compounds should become more stable as they are repeatedly crystallised and their associated morphic fields become more established. A measure of how stable any particular compound is is its melting point, or how high the temperature needs to be before it breaks down. If Sheldrake's theory is right, then these melting points should increase over time, showing that the morphic field is becoming more engrained. This tendency for the melting point for new compounds to increase has indeed been observed by some chemists, although it is usually attributed to other causes, such as improvements in the purity of the compound. In order to examine this question, Sheldrake compared the historical melting points of certain compounds to their current melting points and found that they did indeed seem to increase over time. For example, the melting point of the artificial sweetener saccharine, which was first synthesised in 1878, was 220°C in 1902; in 1996, it was nine degrees higher, at 229°C. A similar effect can be seen for most other novel chemicals. This tendency seems to have been previously unnoticed and cannot be explained by random fluctuations in measurements that may be caused by technological improvements in equipment, because these would likely manifest in both higher and lower melting points and not a continuous increase. Sheldrake's theory of morphic fields provides one explanation for this otherwise apparently inexplicable phenomenon that seems to contradict entirely the conventional wisdom in science that physical behaviours remain constant.[243]

As well as aspects of biology, physics and chemistry, Sheldrake also uses morphic fields to account for attributes of consciousness such as memory and behaviour. It is usually assumed that memory is stored in the brain in particular arrangements of connections between neurons, known as synapses. Neuroscientists have come to this conclusion through observing changes to the structure of synapses that occur when an organism creates a memory.[244] However, Sheldrake and others have pointed out that just because brain structures correlate with memory formation, this does not mean that memories are actually stored there. To use the example of a computer, we can notice changes to the processing unit as it uses information but this is not the source of the information; this information, or memory, is to be found external to the computer, in the cloud. In fact, a particular computer's processing unit (analogous to the brain) can be entirely destroyed but the information it once used (its memory) can still be retrieved from the cloud using an entirely different computing device.

One graphic parallel to this in biology comes from how butterflies and moths have been shown to retain access to memories that were formed when they were a caterpillar, despite having their whole body and brain turn to a kind of goo when undergoing the process of a metamorphosis into a completely different kind of body in a stage of some insects' life known as the pupal stage. Whilst it is assumed that the synaptic connections associated with memories must somehow be preserved in this mush—because brain-based models of consciousness can provide no other explanation—as one study puts it, "this hypothesis has yet to be tested".[245]

Some philosophers have also highlighted a logical problem with the standard model of memory, known as the coding, storage and retrieval model. In order for any particular memory to be retrieved in an information system, whatever is doing the retrieving needs to know something about the memory it is trying to access so that it can recognise and access it. This is known in computer database terminology as using a key to retrieve a particular data entry. This means that the retrieval system itself needs a memory store to hold these keys, which will need its own retrieval system, which will in turn need a memory store, and so on ad infinitum. This is a potentially fatal argument of infinite regress that has seemingly been ignored by neuroscientists who are attempting to understand the mechanisms of how the brain stores memories.[246]

Another issue is how some animals seem to have a kind of memory that persists over generations even without engaging in any conventional learning process. One example is how spiders are able to weave intricate web designs that are unique to their own species and which they have not been taught. Again, the assumption is that this is somehow encoded into their genes, although how this can be achieved for such complex behaviour is difficult to envisage given that the main function of genes is to direct the synthesis of proteins.[247]

Sheldrake's alternative explanation for these types of memory—both for an individual organism and for the behaviours of species—is to suggest that this information is not stored in the brain at all but instead is to be found in non-local morphic fields. The information contained in these fields is retrieved through the resonance of an individual's brain with its own previous brain states, in the case of a single organism, and resonance with a collective pattern established in the species' morphic field created by innumerable generations for species-wide behaviours.

As evidence for this proposition, Sheldrake notes that some spontaneous animal behaviours have been observed being adopted by other members of that species even in circumstances where conventional learning methods like observation are not possible. For example, he describes how blue tits in the south of England discovered how to peck at the foil caps on milk bottles that had been delivered to doorsteps so they could drink some of the contents. This novel behaviour spread rapidly throughout the UK, with some instances being recorded farther afield in the Netherlands, Sweden and Denmark, despite blue tits having very restricted territories that would seem to mean that transmission by imitation over such a large geographical area and at such speed was very unlikely. Sheldrake explains this in terms of a critical number of blue tits having discovered and repeated this novel but rewarding behaviour so that it reinforced particular morphic fields associated with pecking milk bottles, increasing the likelihood that others would express the same behaviour through resonance with their species' morphic fields.[248]

In another example, Sheldrake describes laboratory experiments showing how the rate of learning of groups of rats in how to navigate a water maze increased significantly over a 50-year period even though the rats involved were not descended from trained rats and populations were physically separated across three different continents. Sheldrake suggests that morphic resonance can explain similar questions regarding human behaviour, for example in how children learn new languages so quickly, which is difficult to account for simply in terms of their imitating those around them. The famous linguist Noam Chomsky, among others, has proposed that there seems to be an inherited structure to language.[249] Sheldrake's theory provides a potential host for these structures in the form of morphic fields. It is also well-established that people's scores in IQ tests have risen over the decades despite the tests being just as difficult to complete. There have been various explanations proposed for what is known as the 'Flynn effect' (named after its discoverer James Flynn) such as advances in the general level of education and improvements in health and nutrition, but none has yet been fully accepted. Sheldrake suggests that morphic resonance may offer a more 'natural' explanation.[250] Sheldrake's theory also predicts that it should be easier to complete a crossword puzzle after a period of time when more people have generated the answers and the overall solution has associated with it a stronger morphic field. As we'll see in Chapter 13, such hypotheses offer themselves well to experimental testing.

Sheldrake suggests that organisms are in resonance with morphic fields of consciousness. This resonance is strongest when there is high similarity with the individual or group of individuals whose behaviour and decision-making has contributed to the morphic field. This provides a mechanism for the encoding and storage of personal memories. An individual person would be readily able to access a specific memory, or repeat a learned behaviour, because they are most in resonance with parts of the morphic field to which they themselves have contributed. These fields of consciousness are also present for groups such as families, sports teams, nations and species that have their own hierarchical fields of consciousness with which members of these groups will most strongly resonate and so influence how members of these groups behave. Sheldrake notes that his theory bears some similarity to Carl Jung's notion of the collective unconscious, which Jung suggested is a realm of consciousness shared by all of humanity. Jung suggested that the collective unconscious might also have a hierarchical nature in that it expresses itself more strongly through the similarities of families and cultural groups.[251]

Features of Consciousness

How might these approaches to consciousness suggesting it is somehow field-like address our three features of consciousness of rich personal experience, differing states of consciousness, and determining between conscious and non-conscious entities?

Rich Personal Experience

Similarly to Carr's hyperdimensional theory (Chapter 8), the nature of consciousness itself is not the primary focus of either of these theories. Instead, they attend mostly to how consciousness might operate in the brain and beyond. Hennacy's suggestion is that a non-local consciousness is expressed by the brain and body through electromagnetic fields and the manifestation of holographic biofields. Sheldrake proposes that consciousness pervades non-physical morphic fields, which are shared. (Although in what form these fields might exist, and in what medium, is not specified. They also appear to be conceptually very different from physical fields that are well-established in science, such as gravitational or electromagnetic fields.) The price paid for these theories' adoption of a more functional approach is to leave essentially unexamined perhaps the most defining feature of consciousness, the question of what it feels like to exist and have rich phenomenal experience.

States of Consciousness

Because Hennacy focuses on electromagnetism and the brain, there is a possible means for her theory to explain states of consciousness based on electromagnetic brain waves. The fact that differing frequencies of these waves are associated with different states of consciousness is uncontroversial. Whilst Hennacy does not specifically address this point, we at least have some route consistent with her ideas through which states of consciousness may one day be explained through electromagnetism and non-local consciousness.

However, it is not immediately obvious how Sheldrake's morphic field theory might lend itself to explanations for distinct states of consciousness, again because of its primary focus on the processes of consciousness.

Determining Between Conscious and Non-Conscious Entities

For both of these theories, we might surmise that to be a conscious entity requires life. In Hennacy's case, life enables the expression of biofields from the activity of biophotons and in Sheldrake's case, being alive enables an organism's resonance with morphic fields. Both of these views would seem to rule out the possibility of inanimate objects, including of course artificial intelligence systems, from being conscious since by definition they are not alive and therefore cannot engage with non-local consciousness.

Weird Phenomena

How might these theories of a field-like consciousness account for weird phenomena? Because fields potentially exert influence over large distances, the theories we've met in this chapter might be more naturally conducive to explanations for phenomena that manifest away from our brains and bodies, such as out-of-body experiences, telepathy and psychokinesis.

Consciousness Beyond the Body

The idea that there is an extended consciousness beyond the body is the basis for Hennacy's theory. This model, she suggests, could explain out-of-body experiences, which might be thought of as journeys through consciousness. However, she provides few details of what she believes to be the nature of this wider consciousness. Hennacy is also open to this possibility for the potential persistence of an individual's identity beyond the death of the physical body such as in reincarnation cases,[252] but again it is unclear exactly how her theory might accommodate such expressions of a personal consciousness.

However, Hennacy's theory does provide a possible explanation for savantism, another weird phenomenon where people are able to demonstrate remarkable skills apparently without the need to learn them. Such cases are often associated with brain trauma or illness, like that of Jason Padgett the furniture salesman, who spontaneously developed advanced mathematical fluency after being hit over the head during a violent attack,[253] or Kim Peek whose brain had a large cavity caused by hydrocephalus, but nevertheless had memorised the exact contents of over ten thousand books, among other prodigious talents.[254] Hennacy suggests that these and other cases might occur because there is a kind of collective memory created by all humans that is situated in the wider electromagnetic consciousness in the form of what she terms 'holographic informational fields'. Whilst this is shielded from most people by the magnetite surrounding the brain meaning they usually only have limited knowledge of this extended consciousness, she suggests that savants might be able to more freely access this field because of a disturbance to the magnetite caused by some external disruptive factor such as an illness or trauma.[255]

Sheldrake also argues that it may be possible for consciousness to persist without the body. This is consistent with his suggestion that memories, for example, are not stored within the brain. Whilst this is not the focus of his theory of consciousness residing in morphic fields, he has suggested that death may be like dreaming without being able to wake up.[256] Just as our dreams whilst asleep are informed by our experiences, desires and memories, this may also be the case after our bodies die. This would lead to different accounts of an afterlife depending on our life experiences and expectations, and would explain why Hindus, for example, report different features of near-death experiences than those commonly reported by people in Western cultures. Sometimes these differences relate to how particular aspects of the near-death experience are interpreted. Whilst Western accounts often include an instantaneous and 360-degree life review, in Indian accounts, this sometimes takes the form of Yama, the Hindu Lord of the Dead, inspecting akashic records that contain details of a person's actions during their life.[257] Sheldrake speculates that different theologies and metaphysical viewpoints may result in matching experiences in a dream afterlife; for example, extreme Islamist suicide bombers being rewarded in heaven with 72 virgins and atheists who believe that death will bring an empty void experiencing nothingness.[258]

Consciousness Transcending Space, Time and Matter

Hennacy's observation that ELF waves are produced by the brain in the form of theta brain waves and that these can travel over extremely large distances opens the possibility for explaining how telepathy might work. After all, electromagnetism is how mobile phones and radios work so it seems to make a kind of intuitive sense that perhaps this might be how brains share information with each other. She has conducted research that has identified non-verbal autistic children who appear to be capable of performing telepathic tasks with extremely high levels of accuracy. These children seem particularly effective at doing this with people with whom they are emotionally bonded, such as members of their family or their carers. Her experiments testing for telepathy involve the attempted mental transfer of randomly-selected words and numbers of twelve digits, which should rule out the possibility of getting these tests right through guessing. The set-up includes multiple video cameras and visual barriers between the child and the person 'sending' thoughts to counter any accusations of fraud. Hennacy reports that one autistic child, known as Hayley, was able repeatedly to read the mind of her therapist with an amazing success rate of 100%.[259] This of course goes far beyond the levels of accuracy achieved in most parapsychological experiments on telepathy, such as those described in Chapter 4.[i]

Hennacy suggests that the autism condition itself may be responsible for such seemingly high psychic functioning. She points out that autism is associated with brain functioning that appears to equate with a reduction of the mental chatter that most people experience for most of the time whilst awake. It has long been noted within parapsychology that reducing this waking mental noise can help to facilitate psychic experiences. Some experimental set-ups are designed to do exactly this in order to promote conditions that may be more conducive to weird phenomena such as telepathy or precognition. This includes the ganzfeld, discussed in Chapter 4, which aims to induce in participants a drowsy state through providing a reddish visual field and white noise to reduce stimulation from the normal senses.[260]

Most parapsychologists have long ago dismissed electromagnetism as a means for psychic information transfer between people because successful ESP experiments have been performed with participants in Faraday cages and at large distances, which effectively shields most electromagnetic frequencies. However, there may be a loophole in that such restrictions do not apply to ELF waves, which can pass through most physical structures and can be transmitted over a very wide area. Hennacy speculates that the magnetite surrounding the brain acts like an antenna that detects ELF waves and perhaps also works as a filter for these waves to ensure our thoughts mostly remain private and inaccessible to others (in a similar way to

---

i Although it should be noted that, at time of writing, properly controlled telepathy experiments on autistic children have not been published in any peer-reviewed journal, meaning that exercising a degree of caution regarding these results is for the moment probably advisable.

how only some people, perhaps those with compromised magnetite lining through trauma or illness, have savant skills).

Whilst this all sounds compelling, one potential fly in the ointment with ELF waves is that they are very low resolution and have a small bandwidth for communicating information. Although ELF waves are used for communication between submarines and their base, this requires extremely large transmission sites and they are typically only used for simple messages, such as signalling a need for the vessel to return to the surface to receive more detailed communications by other means.[261] This would appear to limit the potential for ELF waves to be the carrier for psychic information such as might be conveyed using telepathy. However, recent research has shown that complex information could be encoded with minimal bandwidth using quantum information in the form of qubits.[262] Of course, this does not necessarily mean that psychic information is quantum in nature, but it does demonstrate that we should be wary of not allowing our limited current knowledge to blind us to potential mechanisms for psychic transfer that were previously, and perhaps prematurely, discarded.

Like Hennacy, Sheldrake has also conducted experiments in aspects of what he calls the 'extended mind', including numerous scientific tests of various aspects of telepathy. Instead of electromagnetism, Sheldrake proposes that morphic fields might be the basis for the transfer of information from one consciousness to another. He suggests that consciousness is encapsulated in morphic fields that extend beyond our bodies and so, in the right circumstances, the thoughts and feelings of one organism might be accessible by another, perhaps through a morphic resonance between them. He suggests that this means that telepathy should more commonly occur among people and animals who have close emotional bonds because of the resonance of their similarities in the morphic field. This theory has provided the basis of experiments that he has performed on dogs that apparently know when their owners are coming home without recourse to any known physical cues, and telephone telepathy, when people seem to be able to know who out of several close friends or relatives may be phoning them, both of which we discussed in Chapter 4. Sheldrake suggests that telepathy experiments among people or animals who know each other well are more likely to show higher effect sizes than experiments among people who are strangers because there is more resonance between those with close ties, something the results of his experiments seem to show. He also proposes perceptual fields generated by the act of looking that extend outwards from brains that may explain why people seem to be able to 'feel' when someone is staring at them without recourse to normal senses, an effect that has become known as the 'sense of being stared at'.[263]

Free Will

As noted throughout this chapter, Hennacy's theory suggests that the source of consciousness lies beyond the brain and body. This allows for the idea that we—our 'selves'—are the primary agent in control of our own actions and decisions. Whilst her theory does not incorporate a detailed explanation for free will, if our consciousness can exist independently of our bodies, such as in out-of-body experiences, then it should follow that we can act free from any constraints that might be present within our biological and physical make-up. However, she notes that our sense of free will and control over our own bodies can be eroded when we suffer from conditions such as a stroke, which, if it occurs in the right hemisphere of the brain, can lead to patients believing that the left side of their body doesn't belong to them. Another situation that seems to demonstrate that our consciousness and our body are not equivalent is the 'fugue state', which manifests as the experience of a kind of 'autopilot' mode, where the person in question is able to perform highly complex tasks and activities without having any conscious awareness. When a person 'wakes up' from this state, they are frequently bewildered and confused with no memory of how they might have arrived at their current location or what they have been doing. Whilst there is seemingly an absence of free will in these and similar situations, the fact that these are not normal appears to suggest, according to Hennacy, that our having free will is consistent with our default state of being.[264]

Brain-based theories of consciousness are based on a reductionist worldview where causes in nature work from the bottom up and therefore our consciousness must, they maintain, be generated by the activity of the brain. This leads to a perspective of determinism where everything that happens must be described in terms of mechanistic causes, the logical consequence of which is that free will is therefore illusory, as discussed in Chapter 6. Sheldrake's theory is instead predicated on the idea of top-down causes where the mind works through fields. He argues that fields work through such top-down causation. For example, a magnetic field causes patterns to be formed in iron filings; the filings themselves do not cause the pattern. Likewise, according to Sheldrake, our minds work through fields that cause changes in the brain and lead to actions to be performed. So when we decide to move an arm, this is our free choice conveyed to the brain through fields of consciousness. Whilst there are of course constraints to what we can and cannot do in the world due to physical, cultural and environmental limitations, Sheldrake argues that we nevertheless retain the freedom to act within these constraints. He argues, therefore, that free will is not only a commonsense and the 'best scientific' view but that it is a true reflection of reality.[265]

Chapter 10

Consciousness is Information

Over the last three chapters, we've looked at whether consciousness is quantum mechanical, whether it's to be found in higher dimensions of space and time, and whether it's related to fields. In this chapter, we'll discuss theories suggesting that consciousness is in some way related to information. The first of these theories proposes that consciousness arises when information within a system becomes sufficiently integrated (interrelated) and the second that a fundamental all-pervading consciousness uses a form of digitised information technology as a means to evolve. As we shall see, whilst both theories essentially regard consciousness as manifesting in an informational form, their approach is markedly different.

Consciousness is Integrated Information

Integrated information theory (IIT) was first proposed by the neuroscientist and psychiatrist Giulio Tononi in the early 2000s, and has since been developed by a number of other theorists and researchers.[266] The two basic ideas of IIT are this. Firstly, that the fact of consciousness should be self-evidently obvious by the fact of us being aware and therefore conscious experiences must be fundamental to existence. Secondly, that consciousness itself is equivalent to information that has reached a certain threshold of internal integration. This means that consciousness is present in any structure if the information that can be generated by that system as a whole is greater than the information that can be generated by the sum of its individual parts. The most immediate example of such a system is a human brain, although the theory allows for the possibility that other types of biological and non-biological structures to be viable as conscious entities. No matter how extensive, information in itself is not conscious but sufficiently integrated information is. The theory takes the form of a mathematical formulation and makes a number of predictions about how consciousness manifests in the brain and which forms of structures are capable of hosting consciousness and which are not. These predictions should be testable and thereby show how effective this theory is as an explanation of consciousness.

The key contribution of IIT is its suggestion that consciousness is integrated information. The first thing we need to understand this theory is a clear definition of information and how it might relate to consciousness. In order to address this question, Tononi invites us to consider the difference between a task performed by a simple device that responds to light, known as a photodiode, and the same task performed by a human.[267] Photodiodes are basic components of many modern electronic devices and have a wide range of applications in the modern world, including, for example, triggering a street lamp to be turned on or off depending on whether it's night or day. This rather mundane task can also be performed by a human who is asked to report whether it's light or dark outside. The question related to consciousness is what is the difference between the photodiode's (presumed lack of) experience and a person's experience of this same task?

Most people would agree that whilst humans have a subjective experience of light and dark, the simple electronic circuitry of a photodiode does not. According to IIT, one of the key reasons for this is because of how much information each can generate. The photodiode has only two states relating to whether it's light or dark. The information relevant to determining this simple fact can be expressed using a single computer bit, which can either be on if it's light (represented by a one) or off if it's dark (represented by a zero). A human, on the other hand, is capable of determining between a much larger range of alternatives. In relation to how light it is outside, a person can, for example, tell whether it's day or night, whether it's dusk or dawn, or whether it's sunny or we're in a thunderstorm. Therefore, whilst both the human and the photodiode can tell the difference between light and dark, there is a vastly greater amount of information capable of being generated by the human than the photodiode. And, of course, photodiodes, unlike humans, do not even know that they're detecting visual information as distinct from sound or smell or any number of other qualities.

Therefore, it seems that being able to discriminate between many alternatives is important to consciousness. However, the amount of information a system can accommodate is not sufficient for there to be consciousness present. According to IIT, what is also important is how much the information is integrated. In order to understand why this is key to consciousness, Tononi suggests that we consider a digital camera that has a million photodiodes, each of which can discern between light and dark.[268] Such a device is clearly more complex than a single photodiode that can generate information equivalent to just a single bit. The camera with its million photodiodes has a million bits of information which collectively are capable of representing a vast number of possible combinations in the form of simple photographs. ($2^{1,000,000}$ of these different states, to be precise.) We can interpret this camera's information capacity as comparable to the human's in relation to the task at hand in that both it and a human can determine many subtleties of light and dark, such as the difference between when it's sunny with a clear blue sky and cloudy, but still clearly daytime.

Does this much larger capacity for information mean that the camera is conscious? Again, most would say that it probably does not. So what is the difference here between the camera and a conscious human? The key point according to IIT is that the information in the camera is not integrated—it's simply a set of a million individual photodiodes that have no relationship with each other. Conversely, a human being is a complex integrated system incorporating a brain with networks of neurons and regions that interact with each other in a multitude of different ways. Hence the information that can be held in a human brain is qualitatively different from what can be stored as unrelated photodiodes in a camera because the information in a brain is highly integrated whereas the information in a camera is not. When we think about how this feels subjectively, IIT proposes that integrated information means that each of our experiences feels like an integrated whole. For example, if we look at a yellow circle, we don't have an experience of yellow that is separate from our experience of a circle. It's a unified, complete—integrated—experience. This is what it means to experience integrated information as consciousness.

It would of course be possible to build a camera that does have integrated information. Indeed, the cameras in our mobile phones are part of complex computer systems with relatively high degrees of integrated information, especially when we consider these devices' interactions with wider networks of computerised devices on the cloud. So how much information integration is required for any particular system to be conscious? IIT's approach involves two stages.

The first is to calculate the amount of information generated by a system (one bit for a photodiode; a million bits for the camera). The second is to determine how much of the information generated by a system is integrated. This is done by working out how much information is generated by the parts of the system on their own and comparing this with the amount of information generated by the system as a whole. If the second value is larger than the first then, according to IIT, it's conscious. The associated mathematical formula results in an overall value for the system's integrated information that's expressed using the Greek letter phi, which has the symbol Φ. This symbol is particularly apt because the vertical bar can be seen as a capital 'I' for Information with the circle representing its integration.[269]

This is the theory in a nutshell. However, there is a practical issue with this approach when we look to make these calculations for systems that we see in the real world, particularly highly complex systems like brains. Working out the degree of information integration (Φ) of a system involves calculating the probability of its transitioning into each possible end state from each possible starting point. Dealing with this vast quantity of potential combinations and nested combinations very quickly becomes prohibitive even in relatively simple systems. The solution proposed by IIT theorists is to use assumptions and approximations to simplify the calculations involved so that they become more manageable without sacrificing insights that come from these analyses. Developing such techniques is currently an active field of research.[270]

So how does IIT fare when we consider real situations involving aspects of consciousness? One such is the question of different states of consciousness, the second of our two defining features. IIT predicts that information integration should be high when we're in states involving active thought, such as when we're awake or dreaming, and low when we're unconscious, such as in deep sleep or under anaesthetic. Studies inspired by IIT that have measured levels of complexity in terms of brain activity in both conscious and unconscious human participants seem to support this prediction. It also appears that high levels of information integration are associated with areas of the brain that neuroscience has found to be most associated with consciousness. For example, the cerebral cortex has large amounts of interactivity between its various regions and it is well-established that damage to this area results in impaired consciousness. Conversely, the cerebellum, despite having a large number of neurons, has relatively little internal interaction and is not thought to be important to consciousness. This is exactly as IIT would predict.[271]

The conceptualisation of consciousness as integrated information fits well with the particular definition of consciousness proposed by IIT. In Chapter 1 I suggested that we consider consciousness as comprising three defining features. Similarly, IIT also has a proposed framework of consciousness.[272] As well as its starting point that consciousness indisputably exists, something to which we can readily attest simply by noticing that we are currently in the midst of experience, IIT also outlines several other key features of consciousness. These include the suggestion that each experience is intrinsic, in other words, it exists for itself—my experience is my experience and not anyone else's. Another is that an experience has specific qualities that make it this experience and not any other experience. A further feature is that any experience is a whole experience, which can't be separated into parts. For example, if I'm looking in front of me, I'm seeing what I'm seeing and not just the right-hand side of what I'm seeing. Similarly, I can't see anything I can't see, such as what's behind me. (This might feel somewhat pedantic; however, such an exacting formulation is what gives the theory more rigour in defining exactly what is meant by consciousness and so what it is that it is attempting to explain.)

There are other features of consciousness that many experiences seem to involve that are not included in IIT's framework, for example, how most experiences seem to occupy some form of space and take place over a period of time (ideas that are central to Bernard Carr's theory that we looked at in Chapter 8), and that experiences are usually accompanied by a sense of our selves—a feeling that we exist as a part of an experience. However, IIT theorists point out that these and other aspects are not always present in experience, for example when we're in certain altered states of consciousness such as dreaming, deep meditation or under the influence of psychedelics. These states can sometimes invoke experiences that have no sense of space or time and/or might involve a loss of ego. This means that these and other features do not seem to be necessary aspects of consciousness.[273]

One interesting feature of the theory is that each particular state of a conscious information system can be conceptualised as its own unique multidimensional geometric shape. These shapes in information can be categorised into types of shapes that correspond to different sense experiences. So, vision experiences form a group of shapes (think perhaps of dodecahedra that have particular orientations and side lengths that make them distinct from each other) that are different from sound shapes (which perhaps might be analogous to tetrahedra) that are in turn different from smell shapes (that form some other type of high-dimensional polygon) and so on. This might explain why, as Tononi suggests, "blue is certainly different from red ... but clearly it seems even more different from middle C on the oboe".[274]

The uniqueness of our own information systems (our brains) also means that, whilst you and I may construct a very similar experience shape from a shared experience, the huge number of dimensions involved in these experience shapes means that they are vanishingly unlikely to be exactly similar. This could account for why we feel we have more things in common with some people but less with others—because our experience shapes are more similar—and why the information shapes experienced by a bat with its perceptual echo location system or by an octopus with its nine brains are likely to be unimaginable to us, even if they relate to a shared situation such as a dark night in the churchyard or swimming over a coral reef.

So far, most of our examples of IIT look to how this theory relates to consciousness in the brain as opposed to non-conscious entities such as photodiodes or cameras. So what makes IIT a theory where consciousness might extend beyond the brain rather than a brain-based theory, such as those considered in Chapter 2? Firstly, brain-based theories start with the premise that consciousness must be a by-product of physical structures. IIT, on the other hand, begins with the assumption that consciousness is, as Tononi puts it, as fundamental to reality "as mass, charge, or energy".[275] In fact, he suggests that it could even be that conscious entities such as you and me—systems with sufficiently high information integration—"exist in a stronger sense" than physical objects because "there is something it is like to be you, whereas if you just have high mass, charge, or energy, there may be little or nothing it is like to be you".[276]

Secondly, as the philosopher Bernardo Kastrup points out, whilst IIT can be (and often is) interpreted as an explanation of how consciousness arises from physical structures and hence is often presented as a brain-based theory, its adherents expressly avoid such framing.[277] In fact, Ignacio Cea and his colleagues have noted that the most recent formulation of IIT, known as IIT 4.0, "openly rejects the mainstream view that consciousness is generated by the brain, positing instead that consciousness is... primary while the physical domain is just an experience-dependent 'operational' construct to understand consciousness from the scientific, third-person perspective".[278] In other words, consciousness is not derived from anything else and physical entities that do not have sufficient integrated information such as fundamental particles, neurons and everyday objects "do not truly exist" in their own right. Cea suggests that physical objects therefore exist only as structures in consciousness and so are "are ultimately reduced to experiences".[279]

Consciousness Uses Digital Information Technology to Evolve

Another, quite different, theory based on the idea of consciousness-as-information is proposed by physicist Thomas Campbell. Campbell suggests that an all-encompassing consciousness is responsible for manifesting within itself our physical universe as a small part of an unimaginably vast multiverse consisting of both physical and non-physical dimensions. He suggests that this has come about through an evolutionary process that has driven what was a mostly uniform consciousness to become increasingly differentiated as structured information through the workings of what is essentially an enormous digital computer. This computer generates universes that are designed as learning environments in which individual beings can accelerate the evolution of their own consciousness to the benefit of the whole. Progress is achieved through individuals making choices that are aligned with spiritual goals, in particular the realisation of love.[280]

Campbell developed his theory over the course of three decades engaging in intentional out-of-body experiences (OBEs) (discussed in Chapter 3) exploring what he maintains is an extended non-physical reality. His first experience of an exotic altered state of consciousness came whilst at college in the 1970s when he learned the mantra-based Transcendental Meditation (TM), then somewhat of a cultural phenomenon, in part because of its association with celebrities such as The Beatles. Campbell experienced dramatic benefits from practising TM and found that by tapping into its 'point awareness state' he "could solve physics problems, design experiments, analyze research data, and write and de-bug computer code ten times more quickly, and with better results… than… in a normal state of consciousness".[281] After completing his research degree, Campbell found a job applying his maths and physics skills to develop electronic device simulations. His supervisor in this role encouraged him to read the book *Journeys*

Out of the Body[282] and introduced him to its author, Robert Monroe, who happened to live nearby. (We discussed reports of OBEs that Monroe experienced in Chapter 3.) This led Campbell and another of his colleagues, an electrical engineer named Dennis Mennerich, to join a scientific team that Monroe was putting together to develop sound frequency technologies designed to induce voluntary OBEs.[283]

Under Monroe's guidance, the team conducted numerous experiments with these altered states. Campbell writes that the "mental space we practiced in was nonphysical… we were active, wilful, autonomous agents within this larger nonphysical reality. We went places, did things, communicated with nonphysical beings".[284] Campbell became so adept at inducing OBEs that he was able to "live in both… reality systems simultaneously and permanently".[285] The team aimed to conduct tests to confirm the 'operational realness and significance' of these non-physical environments and ensure that, through "demanding measurable, repeatable, objective results",[286] what they were discovering about this wider reality was not simply due to an overactive imagination. One early example of such an experiment involved Campbell and Mennerich simultaneously embarking on an extended OBE. They were situated in different rooms in Monroe's house with microphones hanging from the ceiling above where their bodies lay so that they could record a contemporaneous commentary of what they were each experiencing. After the session, they were able to verify that their accounts of the journey that each had had in non-physical reality were almost identical—they reported meeting up and having the exact same experiences and conversations with each other and with other non-physical beings.

The theory Campbell derived from such explorations is based on two starting assumptions. The first is that our extended reality initially consisted of an undifferentiated and uniform proto-consciousness, a state he calls Absolute Unbounded Oneness. This Oneness had few of the features we might associate with consciousness. There was a complete absence of thought or intelligence and the presence of only a slight and dim awareness. Oneness was nevertheless an entity of consciousness encompassing infinite potential but entirely without order (that is, it had maximum entropy). Whilst Campbell suggests that Oneness was finite but, somewhat paradoxically, unbounded, its source (if any) and whether anything lies outside of it is in practice, he maintains, entirely unknowable.

The theory's second assumption is that reality is governed by a 'fundamental process' of evolution. This is a universal application of the principles described by Darwinian natural selection whereby entities that survive and thrive are the ones that best fit their environment, whilst entities that are unsuited to their environment perish. Beneficial modifications of how an entity is constituted and behaves are naturally enhanced through this process whilst disadvantageous features disappear. The natural inclination of evolution is thereby towards increasing order—a reduction of entropy—something we can observe in the huge varieties and complexities of lifeforms on Earth that are exquisitely well-adapted to their environments. The initial stages of evolution for Oneness began when some random perturbation within it caused it to nudge towards a lower entropy state. Its uniformity became slightly disturbed and hence it started to exhibit signs of subtle differentiation. Campbell explains that it is "as if [a] calm, still ocean… has evolved to the point of making a tiny ripple".[287] Hence, a duality (disturbed/not disturbed) emerged within Oneness that enabled evolution to move consciousness in the direction of more ordered states.

Oneness began to embody basic structure in the form of what Campbell calls 'discrete reality cells'. Each of these cells became able to oscillate between consciousness states such as thought and no-thought, more and less, and bright and dim. This is effectively a binary system that is capable of expressing information. These binary reality cells can affect their neighbours as they change state and in this way consciousness-as-information can be processed, communicated and shared, and memories stored and accessed. These structures essentially reflect the functions of an immense computer formed from consciousness, something Campbell terms 'The Big Computer', although he stresses that this "should not be thought of as the big mainframe [in more modern terms, big datacentre] in the sky… but as a metaphor describing the fundamental nature of the process and functionality (computation, memory, and rules) required to implement our reality".[288]

Campbell calls this new form of lower entropy consciousness that is no longer a uniform Oneness, Absolute Unbounded Manifold (AUM).[i] Through the evolutionary process of differentiation, individuation and specialisation, AUM becomes increasingly complex. In

---

i Perhaps not coincidentally, AUM is how OM is sometimes spelt. In Indian philosophies such as Hinduism and Buddhism, AUM/OM is said to be the primordial sound of the universe.

order to accelerate its evolution, AUM begins to create what might be thought of as distinct compartments within consciousness in the form of alternative realities or 'dimensions'. These dimensions offer opportunities for individuated units within AUM to interact with other individuated units (essentially AUM interacting with itself) in a multitude of different ways in experiments designed to actuate the most productive dimensions that promote speedy evolution. Dimensions are formed by The Big Computer generating subprograms (somewhat akin to mobile phone apps), each having its own distinct set of rules, capabilities and contents. Because of the process of evolution, the dimensions and encapsulated individuated units within them that are the best suited to reducing entropy—more ordered consciousness—survive, thrive and contribute to the evolution of the whole, whilst less productive dimensions and units that increase entropy—more disordered consciousness—decline and disappear.

Individuated units of consciousness that inhabit dimensions are essentially living beings. Each dimension in which they operate is a unique learning environment designed for the purpose of enhancing the quality of consciousness for the individuals who live there for the ultimate benefit of AUM as a whole. The first dimensions to be created were non-physical in that they featured relatively few constraints for the beings who lived in them. These are fluid places where thought is primary and there is no time or space, at least as we experience these concepts. Whilst evolution of consciousness does take place in non-physical dimensions, it was found that they are not well-suited to the evolution of basic, elementary, less evolved units of consciousness. Campbell suggests that this is because it can be difficult to pin down which individuals caused particular actions and their consequences due to the relatively few constraining rules and lack of solidity in the environment. This results in a slower and more erratic pace of evolution.

To address this issue, new dimensions were created to embody more consistent sets of constraints and rules. One example of this type of dimension is the physical universe in which we live. Physical dimensions are highly structured in order to promote the more efficient evolution of individual expressions of consciousness such as ourselves and the other beings with whom we share our world. Many individuals in these dimensions represent relatively low states of consciousness organisation. (With what might be thought of as a trademark sense of humour that suffuses his books, Campbell comments, "if you ever wondered why a nice digital-consciousness-love-being like AUM would populate our world with so many jerks, now you know".)[289] Because these dimensions are specifically designed to support the learning of basic units of consciousness, they might be thought of as performing a similar role to the optimal learning environments we create for children where clear boundaries are established to support healthy psychological development.

In the case of our dimension, these constraints in consciousness are what we interpret as the laws of nature. The features of the rule-set governing our dimension are what we refer to variously as the physical properties of mass, energy and forces, derived and measured through science, particularly the disciplines of physics and chemistry. Time is created through the regular oscillation of reality cells from one binary state to another. It is a tool to enable the separation of past from future and has advantages for accelerating the evolution of consciousness because "complex interaction and causal logic chains [can] evolve as the beat goes on. Entropy is reduced. Time acts as a catalytic agent. It dramatically enhances [AUM's] ability to self-organize".[290] The notion of space is derived from time because of the way information is passed from one reality cell to another. The maximum speed of this propagation of information in physical spacetime is the speed of light as described by Einstein in his theory of special relativity. Campbell suggests that his model of reality with its discrete consciousness cells explains why space is granular or pixellated at Planck scales (thought to be the smallest units of space) and why the speed of light is the same for all observers.

It's important to bear in mind that, despite its apparent physicality, our dimension, in common with the rest of AUM, is comprised entirely of consciousness. This is consciousness governed by strict rules and constraints in order to deliver the experience of a solid and (mostly) predictable material reality. Space is not some substance that is separate to consciousness. It is formed from consciousness and suffused with a set of constraints that are unique to our dimension. As Campbell puts it:

> Space is an illusion, a trick of your individuated mind, a virtual reality within the actual reality of your consciousness. Space, as we experience it, is not fundamental. It is an attribute or property of our virtual reality that is generated by its defining rule-set.[291]

Whilst Campbell derived his theory of consciousness and physical reality through his explorations in consciousness, he notes that it bears similarities with tentative ideas proposed by leading physicists of the twentieth century, which were developed using more conventional means. For example, he writes:

> Albert Einstein (Unified Field Theory) asserted a nonphysical field as the basis for both matter specifically and reality in general, thereby moving science closer to the truth, but he did not appreciate the discrete digital nature of both space and time or the role of consciousness (instead of space-time) as the primary energy field. Einstein's student and colleague, the great quantum physicist David Bohm (along with a few of the best quantum mechanics theorists such as Niels Bohr, Werner Heisenberg, and Eugene Wigner) made the consciousness connection but missed the digital connection.[292]

Our role as individuals within the learning environment of our physical reality, in common with all other individuals in dimensions both physical and non-physical, is to evolve the quality of our particular parcel of consciousness by decreasing its overall entropy. This is essentially a process of spiritual growth involving beneficial and productive decision-making and experiencing the consequences of our actions, a process known in Indian philosophies as the law of karma. Opting for the most judicious choices in our lives can be supplemented with spiritual practices such as meditation, journeys into non-physical realities through OBEs and/or assistance from non-physical beings (sometimes experienced as intuition or insight). Campbell sums up spiritual growth as "changing your attitude, expanding your awareness, outgrowing your fears, reducing your ego, and improving your capacity to love".[293]

Campbell suggests that there are no guarantees of success for any individual. Our choices and behaviour matter. The process of evolution is neutral and segments of consciousness that prove unsuitable to their environment will naturally become extinct. Campbell notes that humanity is itself in danger of this fate as collectively we are in the process of potentially wiping out much of life on Earth through self-destructive activities such as proliferating nuclear weapons and wars, causing climate change, degrading biodiversity, and engaging in unsustainable resource use and pollution. As such, he writes:

> Humanity, the master of its physical world, must evolve a commensurate quality of consciousness… If we do not evolve the whole of our being, if we do not achieve sufficient balance, humanity will become self-eliminating—at least to the point where the remaining dysfunction no longer jeopardizes the whole. Nothing personal.[294]

Features of Consciousness

How might these informational approaches to consciousness address our three features of consciousness of rich personal experience, differing states of consciousness and determining between conscious and non-conscious entities?

Rich Personal Experience

Although very different, both IIT and Campbell's theory propose that consciousness comes first, that it is fundamental to the universe. In the case of IIT, this consciousness finds expression when a system's information is sufficiently integrated. In Campbell's theory, consciousness is all there is. Just as we see with other theories in this book suggesting that consciousness is fundamental, this means that there is (again arguably) no need to explain how consciousness feels like rich experience—this is just its essential nature.

States of Consciousness

In a similar way to brain-based approaches and Hameroff and Penrose's theory, we might surmise that IIT should in principle be able to account for states of consciousness with reference to distinct configurations of the feature it suggests is key to consciousness, namely a system's integrated information. However, much like these other approaches, the focus of research in IIT has so far been on accounting for how there might be a basic presence or absence of consciousness rather than explaining a wider variety of states of consciousness. Hence, in IIT, there is a suggestion that a lack of consciousness occurs when the underlying structure that is said to be the host for consciousness loses its integration. More sophisticated research regarding how IIT might account for more diverse states of consciousness is still in its infancy or yet to be considered.[295]

Whilst Campbell has made extensive use of altered states of consciousness to develop his theory, particularly his explorations of other dimensions using out-of-body experiences, it is not entirely clear how his theory might account specifically for how different states of consciousness might occur.

Determining Between Conscious and Non-Conscious Entities

Of all the theories considered in this book, IIT has perhaps the most to say on how we might determine between conscious and non-conscious entities. Indeed, this issue is its unique selling point, one might say. According to IIT, a system is said to be conscious if the amount of integrated information generated by that system as a whole exceeds the amount of information generated by its parts. In this way, IIT offers a formulaic means for determining between conscious and non-conscious entities. In principle, then, it should be possible to work out whether any system or entity, including any particular artificial intelligence (AI) system, is conscious or otherwise by applying IIT's mathematical schema, although in practice this is challenging due to the complexities involved in such calculations.

In Campbell's theory, everything is ultimately composed of consciousness. The demarcation for Campbell between a conscious and a non-conscious entity is according to whether it has agency in the world and is able to self-organise. Different beings have different extents of free will and self-organisation. Most animals, for example, have fewer degrees of freedom than humans, although this does not mean that they are any less conscious. Campbell regards entities that do not have agency, such as physical objects, as constraints within consciousness rather than as individual beings. Such entities are made of consciousness but they are not themselves conscious because they have no capability for self-organisation. According to this logic, Campbell believes that it is possible for AI to be conscious if such systems have the capacity to act freely and to self-organise. However, he writes, "rather than say that the computer is conscious, it is more accurate to say that the computer supports the natural and automatic formation of consciousness through self-organization".[296]

Weird Phenomena

Whilst each of the theories discussed in this chapter is based on forms of consciousness-as-information, their explanations (or, as we shall see in some cases, dismissal) of weird phenomena are very different. Let's examine each of IIT and Campbell's theory in turn for how they address the questions of consciousness potentially existing beyond the body, consciousness transcending space, time and matter, and living entities having genuine free will.

Consciousness Beyond the Body

The standard interpretation of IIT is that integrated information relies on a physical substrate upon which consciousness can exist. Whilst this may not necessarily involve a biological substrate in the form of a brain—consciousness could conceivably be hosted by sufficiently integrated electronic circuitry for example—it is difficult to reconcile IIT with a consciousness that can survive beyond a foundation of some form of functioning physical machinery.[ii] If the brain decays then it follows that consciousness also withers. As Tononi puts it, "when the brain is shattered, the mind shivers into dim fragments... when the cortex shrinks, consciousness is bound to wilt".[297]

Likewise, from this perspective, out-of-body experiences (OBEs) are entirely mental constructs. Tononi has imagined the great scientist Galileo having an OBE as part of a vivid dream:

> Hovering in midair, he slid inside his erstwhile room—a soul floating above his bed—and the soul saw himself: his eyes were closed, his mouth half open—it was an old man's face. And yet the soul was light, unfastened from the frame of his own body.[298]

The key here is that Galileo is dreaming. The experience is merely a product of Galileo's mind and has no tangible reality beyond his brain. Tononi does not consider that actual journeys of a consciousness untethered from its physical substrate can happen, instead dismissing "both flying dreams and out-of-body experiences [as] due to insufficient blood to a region of the brain underneath the temple".[299] As for reincarnation, this too Tononi regards as impossible because the death of the body entails "the loss of consciousness forever... when information [is] torn to pieces in the brain—disintegration of the body [is] disintegration of the soul".[300]

In Campbell's theory, consciousness is primary and is therefore not limited to the body. In fact he suggests that "there is no physical body, except in your mind"[301] and hence out-of-body experiences (OBEs) and reincarnation become possible, even likely. He regards travelling outside of one's body as simply a change of perspective. As we discussed, the use of OBEs as a means to discover the features of a much wider reality beyond our physical world is the basis for how Campbell formulated his schema. Indeed, he recommends that all us might engage in OBEs to support our spiritual growth. He writes:

> The possibilities for interaction within the larger reality... range many orders of magnitude beyond your wildest imagination, and there are few constraints placed upon what you can do and where you can go. Your aware intentional experiences in [non-physical reality] often

---

ii That said, a potential non-standard interpretation in which high-dimensional 'experience shapes' can exist in non-physical substrates is proposed in Chapter 13.

> become an integral part of an accelerated growth path... Restricting yourself to [physical reality] is analogous to never leaving your house ... fine (and relatively safe) for an infant but rather limiting for an adult. Growing up is what your existence is all about.[302]

According to Campbell, reincarnation into multiple physical bodies over the course of many physical lifetimes occurs in service of our spiritual purpose to improve the quality of consciousness. He suggests that "because of our relatively high entropy, we are slow learners; consequently, our cycles of learning and growth are designed to accumulate results over multiple experience packets".[303] Reincarnation is also not restricted to an experience of being human. If our non-physical 'larger consciousness entity' or 'oversoul' decides in between physical lives that it would be beneficial for the evolution of our individual consciousness, we can choose instead to incarnate in the form of an animal's body. However, because, according to Campbell, most animals have a more limited capacity for free will than humans, such an option restricts our potential for spiritual growth. As he puts it, choosing a non-human body for our next lifetime makes it "more difficult to run up a high quality of consciousness score, but it is also more difficult to lose quality points as well. That is a trade-off you will need to consider".[304]

Consciousness Transcending Space, Time and Matter

Just as a straightforward account of IIT cannot accommodate consciousness existing beyond the body and brain, IIT likewise rejects the notion that consciousness might be able to transcend space, time and matter in the form of telepathy, precognition or psychokinesis. As Tononi puts it bluntly, "there is no magic anywhere: all is mechanical".[305]

Conversely, Campbell regards the paranormal as a 'natural artifact' of consciousness and he suggests that such weird phenomena occur frequently, especially to those individuals with a low entropy, that is, a more evolved or more spiritual consciousness. However, these phenomena are 'largely constrained' because by definition they break our dimension's rules of reality. The whole point of physical dimensions is to provide us with rigid structures within consciousness and so the widespread incidence of weird phenomena would be counterproductive to our evolutionary purpose and growth. As Campbell puts it, "the effectiveness of the learning lab would quickly degenerate".[306] This results in what he calls an 'uncertainty principle' regarding paranormal events which is why the scientific measurement of such phenomena is rarely, if ever, clearcut. Whilst personal paranormal experiences can often be dramatic and entirely convincing to those involved, the rules of our physical dimension mean that there is always some reason why such effects remain elusive to those who would seek to provide definitive evidence of their existence. Nevertheless, Campbell suggests that they occur because they can be useful spurs to help consciousness evolve as they offer signs of a deeper reality or provide intuitive insights through synchronicities.

The instances we see of consciousness transcending space, time and matter are made possible because of the ultimate unity of consciousness (AUM). As Campbell explains it, "all beings, and all things, are on the network, have potential access to each other, and can exchange information as well as other forms of energy".[307] This fundamental interconnectedness of all things, ultimately parts of a singular consciousness, is what accounts for the effects we describe as telepathy, psychokinesis and energy healing.

Precognition has a somewhat different explanation for which we need to provide further details about the role of The Big Computer. According to Campbell, time unfolds through The Big Computer running simulations of every possible scenario that might happen subsequent to any instant, associating each of these futures with a probability, and then storing what happens and what didn't happen as actual and alternative histories.[iii] A precognitive event occurs when an individual accesses one of these simulated possible futures. However, this is only a probabilistic future and so it is possible that it might not come to pass, or its details could change, depending on choices made by the entities involved at each of the instants leading up to moment in question. Campbell suggest that this explains why events in the far future are usually associated with more uncertainty and therefore are more difficult to precognise. This perhaps also might explain the relatively consistent results we saw from presenti-

---

iii This of course is an immense computational task necessitating what for us is an almost unimaginably rapid speed of processing and vast capacity for storing information. However, it is important to remember that these computations take place in the medium of consciousness beyond the constraints of space and time, and so the power of The Big Computer is literally incomparable to the performance of our Earthly computers with their strict physical limitations. Although the power of quantum computing, with its utilisation of features closer to ultimate reality, perhaps provides a hint of what is conceivable.

ment experiments, which relate to futures only seconds away (Chapter 4). Somewhat incidentally, the storing of our personal history and alternative histories as part of The Big Computer's database could also provide an explanation for the life review that some people have reported happening during near-death experiences.[308]

Free Will

Unlike Tononi's rejection of other weird phenomena that are considered in this book, he and his collaborators suggest that IIT does support the idea that we have a free will that is genuine, where "we have true alternatives, we make true decisions, and we—not our neurons or atoms—are the true cause of our willed actions and bear true responsibility for them".[309] The reality of free will from the perspective of IIT, which they point out is in alignment with both commonsense and most religious and spiritual traditions, is based on the starting point of the theory that consciousness exists, a fact we know for certain is true because we ourselves have conscious experiences.

They describe free will as incorporating distinct stages of thinking —imagining alternative choices, choosing one of these courses of action based on some reason or other and then acting on this decision. These stages are all manifested by geometric information experience shapes within consciousness. Whilst it is the case that these experience shapes are realised by the neurons in the brain in what they call the 'operational basis' for the integrated information, the most fundamental reality is the shape itself, not its neuronal foundation. Hence it is consciousness that is in control and can enact decisions, and not the physical substrate which is its host. Further, our freely-chosen actions and decisions within consciousness (that is, integrated information) affect the physical structures of our brain through what has become known as 'neuroplasticity', showing the primacy of consciousness over its substrate and the effects of true free will. Ultimately, the choices we make in our lives bring about changes not only to the physical structure of our brain but also to the world.

For Campbell, free will is a fundamental aspect of his theory, in that the freedom to choose our future is a necessary and innate feature of our consciousness, within, of course, the constraints of the dimension in which we live. So-called natural laws such as causality and determinism that are suggested as arguments against free will only exist within physical dimensions. Consciousness, of which we are made, is not bound by these rules. Indeed, the whole notion of spiritual growth towards ever more ordered states of consciousness only makes sense if individuals are free to make beneficial, or, indeed, harmful, choices. As Campbell puts it:

> The physical causal world is simply a theatrical set (a playground with rules) provided to you to help you unfold your personal drama by forcing you to make significant subjective choices based upon your intent, not objective choices based upon causal logic.[310]

Chapter 11

Consciousness is All There Is

We have so far examined theories that propose that consciousness is a result of brain processes, that it is quantum mechanical, to be found in higher dimensions, has field-like properties and that it is informational. We now turn to our final group of theories suggesting that consciousness extends beyond the brain. The two theories we'll meet in this chapter propose that consciousness is not some accidental by-product of the brain but is instead fundamental to all that exists, that it is the source of all that we perceive and experience.[i] These theories suggest not that our minds emerge somehow from matter but that consciousness is responsible for creating the material world we see around us.

Conscious Life Creates the Universe

Biocentrism is a theory proposed by the biologist Robert Lanza, theoretical physicist Matej Pavšič and astronomer Bob Berman.[311] It builds on the premise that we are the creators of our universe and not the other way around. The theory suggests that the physical universe arises as a direct result of our consciousness – that we quite literally manifest the world through our observations. Whilst this might seem strange from the perspective of how we normally think of our place in the universe, biocentrism is said by its proponents to be the most straightforward perspective we can take from over a century of radical scientific discovery, in particular from the theories that are

---

i Note that this is also what Thomas Campbell's theory proposes (Chapter 10), making this too effectively a theory suggesting that 'Consciousness is All There Is'.

currently our best models of physical reality—relativity and quantum mechanics. These theories together shatter the supposedly common-sense notion that there is an objective space and time that can in any way exist separately and independently from us. In fact, modern physics suggests that spacetime is entirely illusory and, according to the proponents of biocentrism, space and time are ultimately just "modes of interpretation and understanding"[312] that have been hard-wired into the way we look at the world.

Let's dig a little deeper into what this means in relation to the concept of space. We're so engrained in our ways of thinking that it's difficult not to consider space as some kind of container in which we find ourselves and other objects. But that's not really what space is. Space isn't somehow separate from what we find within it—space is an indivisible aspect of our physical universe. Just consider if we were somehow to remove all material objects—all the stars, planets, rocks, gases, lifeforms—what would be left over? Where and what would space be then?

Space is also not constant as we might believe. Einstein showed in his theory of relativity that distances between objects shrink at speeds close to the speed of light. For example, imagine a field that is around one hundred metres long, about the size of a football pitch. If we were to traverse its length at 99% of the speed of light, it would become only fifteen metres long. This is not some illusion—the field would actually *be* that short. And if we were to travel at 99.9999999% of the speed of light, then there would be just four millimetres between each end. From the perspective of a photon—a particle of light, which of course travels at lightspeed—spatial distances vanish completely—everything in the universe is literally in the same place. This is what Einstein meant by space being relative. Space does not have any inherent objective reality. Its structure depends on how it is observed.

We are also used to thinking of objects as separate from each other. However, what we perceive of as boundaries at the human scale become indistinct at the level of the subatomic, meaning that the idea of separateness in space is essentially just convention. There's a chair here, a table over there. This is just a linguistic convenience to facilitate everyday life, which doesn't really bear close scrutiny. We regard an object as distinct when we name it. For example, Lanza and his colleagues suggest we consider the tines of a fork. Each prong is seen as a specific thing only because it has a label. However, we don't have a name for the curved bit of the fork between its handle and the tines so we don't view this as a distinct entity. Whether we call a table 'pieces of wood glued together' or a 'table' is just a matter of perspective. This lack of any real objective boundary between what we might regard as separate becomes clearer when we think of something like a waterfall. You might know where to look when I say 'isn't that waterfall amazing?' but its component parts are forever changing; for example, the water flowing into and out of it. Where it starts and ends is impossible to define exactly.

As with space, the idea that time has some objective reality is enmeshed in our ways of thinking and acting in the world. But Einstein's relativity didn't just sound the death knell for objective space, his theory quashed the idea of objective time too. At extremely high speeds, not only do spatial distances shrink drastically, but time also proceeds more slowly. If you were to travel at 99% of the speed of light into outer space, for instance, then time for you would pass seven times more slowly than it does for those left on Earth. If you made what felt to you like a decade-long interstellar trip, your friends and family would be seventy years older on your return (those who are still alive of course).

But if time has no external reality, what is it that a clock is measuring? A clock is essentially just a way of displaying the passage of some regular process, such as the sun crossing the sky for a sundial, the unwinding of a piece of metal for a grandfather clock, the vibration of a piece of quartz for a wristwatch, or the reliable decay of radiation from caesium for an atomic clock. The purpose of a clock is to provide a reliably-occurring event that is relative to whatever event or activity we want to measure. It helps us to refer to clock time if we want to meet a friend for coffee, say, or see how fast we've just run around the park. But when we use a clock, we're really just comparing the passage of two events—there's nothing independent of any sequence of events that is some tangible entity we can definitively label as time.

This notion is supported by the fact that none of the basic scientific equations for the laws of physics involves time. Time is simply not relevant when describing the most fundamental principles of how the physical world works. This is somewhat surprising given the ubiquity of time in our thinking and in our everyday life. That said, if you've read any scientific discussion of time before, you have probably come across the suggestion that there is one scientific law that may be indicative of some objective notion of time. This is the second law of thermodynamics.

The second law of thermodynamics relates to entropy and it states that everything will tend towards disorder. Whilst order might increase in a particular locality—for example, in the form of a house that's being built—this will entail an overall decrease in order more generally through the extraction of raw materials and in the use of energy. Entropy can be seen in how a cup (ordered) that is dropped onto a hard surface smashes into many pieces (disordered). This is a one-way process and hence seen as providing a concept of time flowing from the past to the future by way of the present. In the past, I had a cup of coffee. Now, I have a mess on the floor. In the future, I'll need to clean it up. It's hard to imagine this process going in the other direction. Superficially at least then, entropy seems to involve an objective direction to time that reflects how it feels to us.

However, this is only one way of looking at entropy. Another perspective (which was indeed the view of Ludwig Boltzmann, the scientist who devised the second law of thermodynamics) is that us seeing ever-increasing entropy in the world is simply a result of so-called ordered states being much less likely to occur than disordered states. For example, if we set up a box with a divider in it with nitrogen in one half and oxygen in the other and then we remove the divider, we'll gradually get a mix of the gases. This is entropy in action. But it's not necessarily an indicator of time, it just shows that the gases are mixing. Each state of the gases is equally likely to occur. Calling the initial state 'ordered' is just us imposing our human subjectivity onto something that is not innately differentiated in Nature.

The bottom line is that space and time, which we think of as fundamental to our world, are really just chimeras, the reality of which vanish on close inspection. As such, instead of consciousness arising from the physical universe, which, as we've seen, the theory suggests doesn't really exist in any objective sense, biocentrism has it that it is us as biological lifeforms who create the world.

There are two aspects to this claim, the first of which relates to how we perceive the universe and the second to quantum mechanics. Regarding the first, there is an age-old philosophical question that goes 'if no one is in the forest to hear it, does a falling tree make a sound?' Most people will answer this with a resounding 'of course' as if the alternative is ridiculous. However, on considered reflection, we find that the correct answer should be 'no'—there is no sound without us. Here's why. A falling tree generates pulses in air pressure. If a person is nearby and the pulses fall within a particular range, then their eardrum will vibrate and send an electrical signal to the brain, which will be interpreted as sound. This is what sound is. If no one is nearby, or the frequency of the pulses falls outside of our range of hearing, then the falling tree literally has no sound. Another example might be a rainbow caused by a garden lawn sprinkler in the sunshine. This rainbow only exists for someone standing in the right place to see a refraction of light. If no one is at this location, there is no rainbow. We quite literally create the sound of falling trees and we create rainbows.

Support for the second aspect of us creating the world comes from quantum mechanics. As we saw in Chapter 7, this world of the very small is strange indeed, with (for example) particles that can be in more than one place at the same time and entangled particles able to transfer information across light years instantaneously. Countless experiments involving ingenious variations on the famous quantum double-slit experiment have shown that it is how we observe subatomic particles that seems to determine what they do. This applies not just in real time—how we look in the present also affects what actions these particles performed in the past. Some experiments have shown that the state of photons can be influenced by events that have not yet occurred. As we also discussed in Chapter 7, interpreting quantum phenomena as the result of a conscious observer collapsing the wave function from a cloud of possibilities to a definite material reality is known as the Copenhagen interpretation. Whilst it is not the only interpretation of quantum mechanics, it is according to biocentrism the most straightforward explanation for the effects we see in all experiments conducted so far. Quantum mechanics in the form of the Copenhagen interpretation therefore shows us once again not only that space and time do not exist in any real sense but that we as conscious observers are able literally to create physical reality.

As also discussed earlier, a common approach for dealing with the mind-blowing implications of quantum mechanics is to suggest that the findings of its experiments apply only to the subatomic world. According to this view, quantum mechanics does not apply to human-sized scales. Our everyday world is assumed to operate according to classical physics, in other words, entirely predictably along common-sense lines without any weird quantum effects and with some objective existence without us. However, recent research is blowing a hole in this comfort blanket. In the last decade or so, scientists have demonstrated that quantum superposition (effectively being in two places at once) can take place at a distance of half a metre and over a timescale of one second,[313] and that living organisms in the form of bacteria can be manipulated to be quantumly-entangled with photons.[314] This shows that quantum effects most definitely can happen in biological life and at scales we are used to.

If reality does not exist until we observe it then what does this say about objective reality? Is my reality the same as your reality? In the 1960s, the physicist Eugene Wigner conducted a thought experiment in which he would make a measurement in his lab on a particle in a state of superposition—that is, it could be in any number of places—but which collapses to a single definite location when he makes his observation according to the laws of quantum physics. What might this mean for Wigner's friend who is outside the lab who does not yet know the location of the particle that Wigner has observed? Is it still in superposition for this friend or has Wigner's observation collapsed it to the same state for everyone? It would seem to boggle the mind that Wigner's friend might experience a different reality from Wigner. However, scientists have now been able to test this hypothetical scenario. In an experiment, they constructed a quantum system involving photons existing in two versions of reality that are *incompatible*. Their results demonstrated that observers can indeed experience different competing realities, no matter how much that seems to contradict our assumptions about the world. The researchers who conducted this experiment commented that their findings seem to undermine the very basis of the scientific method, which of course relies on an agreed external reality.[315] It also brings into question how we view normal daily life where we assume that everyone around us is experiencing more or less the same physical world as we do. Perhaps this is simply not the case and we (literally) create at least some aspects of our own universe.

The idea that our consciousness creates the universe through observation may provide an answer for why the universe appears to be so exquisitely finely-tuned for our existence. It has been noted by scientists that if many of the laws of nature were only slightly tweaked, this would mean that humans could not exist.[316] For example, if the universe had expanded just one billionth faster than it did, then galaxies and stars would not have had enough time to form. If the strong nuclear force that holds atoms together was just 2% less, then the only element that could exist would be hydrogen. If carbon had only a slightly different resonance value, then it would be much less abundant and carbon-based lifeforms would be impossible. There are tens, if not hundreds, of similar examples of constants in physics requiring values to be within a very small range to allow for our existence. Any slight deviation in any of these would result in us simply not being here.

A common explanation for this extraordinary fine-tuning of the universe is the Anthropic Principle. This states that if any of these values was different, then we wouldn't exist, so we couldn't be here pondering the question. Because we clearly are here, then this is how the universe must be so it's actually not so strange after all. Lanza and Berman dismiss this as "circular reasoning or a facile way of squirming out of explaining the enormous peculiarities of the physical universe".[317] Instead, biocentrism supports what the physicist John Wheeler termed as the Participatory Anthropic Principle where observers create the universe, just as biocentrism suggests. This creation process necessarily involves all of the fundamental forces and constants to be in a state that allows for such an observer. They suggest that this is a more straightforward and parsimonious explanation for the finely-tuned universe.

Remarkably, the full implications of this viewpoint mean that our observations are not only responsible for creating the here and now but also for creating a complete, coherent and consistent history that is compatible with us being here in the present. Without an observer, the universe would remain in an unresolved state of possibilities. When we create the universe through our observations, the universe we see is necessarily made compatible with our existence. We create our present reality but we also create its past. According to biocentrism, we are the first cause and the creators of the particular previous configurations of everything that has led to us being right here, right now—from the big bang to geology to biological evolution.

Conscious Agents Outside of Space and Time

As with biocentrism, the cognitive psychologist Donald Hoffman has also developed a model of reality that does not start with any notion of physical space and time, but instead takes consciousness as fundamental and attempts to show how spacetime arises from that. He suggests that there are entities beyond space and time which he calls 'conscious agents', individual units of consciousnesses that are ultimately part of a single, all-encompassing consciousness. It is the interactions of these conscious agents that give rise both to the physical world and to what we experience as our minds. Through developing mathematical models of conscious agents, Hoffman believes that he and his colleagues will eventually be able to explain aspects of the physical world that we currently assume to be fundamental, such as the speed of light, thus demonstrating conclusively that the universe arises from consciousness and not the other way around.

Hoffman proposes that the physical world emerges from a consciousness that is outside of space and time. He suggests that this consciousness is made up of networks of conscious agents interacting with each other. These conscious agents can have experiences, they can perceive other conscious agents, they can make decisions and they can perform actions. Conscious agents are not only the basis for our own individual consciousness and the consciousness of other organisms, but they also give rise to inanimate objects such as rocks, molecules, atoms and the so-called fundamental particles such as electrons, neutrinos and bosons.

Hoffman aims to develop in detail the logical and mathematical principles that underpin fundamental consciousness beginning with just two basic assumptions related to the operations of these conscious agents—their experiences and their actions. He suggests that these two assumptions are enough to lead to a mathematical explanation for the entirety of the physical world and spacetime, including all physical laws, the immense diversity of life and the wealth of human experience including concepts such as memory, the self and ethics.[318] His schema is based on mathematical formulations known as Markovian dynamics, a branch of mathematics that deals with the evolution of states, or configurations, of entities.[319]

Central to Hoffman's approach is the assertion that what we perceive as real with our senses is not the world as it is but is instead merely a representation of reality that bears little relationship to what truly exists. In overtones of the film *The Matrix*,[320] in which reality is said to be generated by a massive computer, Hoffman suggests that we interact with a reality that is wholly virtual. He likens this to our experience of working on a computer desktop interface. We know that the files and icons and text that make up such a desktop interface are not actually the underlying computer. They are instead just useful representations of its electronic circuitry. It would be next to impossible for us to interact directly with the inner workings of any computer. Our productivity would shudder to a halt and we would find ourselves unable to do anything useful such as write an email or watch a video. In the same way, objects in the world that we appear to see and hear and feel do not actually exist in any objective sense but instead are merely useful representations of the underlying objective reality. Hoffman writes that "ignorance of reality can aid command of reality. This claim, out of context, is counterintuitive. But for an interface it's obvious".[321] He and colleagues call this the Interface Theory of Perception.[322]

Hoffman suggests that we have evolved to perceive only what is useful to us in the world (which he calls 'fitness') and not what is actually there ('truth'). Through a process of modelling and simulation, he and his colleagues have proposed a theorem called Fitness-Beats-Truth.[323] This, they have shown, demonstrates that organisms that perceive fitness in the environment will drive to extinction organisms that perceive truth. In other words, organisms that are able to perceive what helps them best to survive in their environment will be much more likely to survive and produce offspring than a similar organism that is tuned instead to perceive the true nature of reality. Whilst on first inspection this seems surprising —surely we see the world just as it is?—a little introspection can reveal that this is exactly how we operate in the world. For example, we can't directly perceive oxygen using our senses despite its absolute necessity to our survival—in fact, it was only discovered in the eighteenth century. Whilst for most of our evolution we literally did not know of the existence of oxygen, we were of course able to tell whether the atmosphere we're breathing might be problematic—too little oxygen makes our head ache, too much makes us faint. Natural selection has provided the tools we need to respond appropriately to our environment—to move towards what helps us and to move away from anything that harms us—but not to perceive it with accuracy. Hoffman says that, according to his theorem,

> mathematically it turns out that the probability is zero that any sensory system has ever been shaped by natural selection to see any true structure of objective reality... Evolution didn't shape us to see the truth; evolution shaped us to have a user interface that helped us to stay alive.[324]

As with biocentrism, Hoffman goes beyond claiming that it is simply our perceptions of reality that are wrong. He maintains that the very objects we see in our world, and even the fabric of spacetime itself, do not exist in any real sense. Our best theories of physics show us that our direct experience of reality is illusory. In our discussion of biocentrism, we saw that space and time are very different depending on the circumstances of whoever it is that is observing them. It is also the case that Einstein's theory of reality breaks down below the thresholds of 10^{-33} cm in distance and 10^{-43} seconds in time. This theory—the best we've got to describe space and time—simply ceases to apply at very small scales. Hoffman quotes one leading physicist, Nathan Seiberg of the Institute for Advanced Study at Princeton, who states that "I am almost certain that space and time are illusions. These are primitive notions that will be replaced by something more sophisticated".[325]

If physical objects and space and time are not the basis for reality, then what is? Hoffman maintains that our four-dimensional spacetime (three dimensions of space plus one of time) is in fact a structure of two-dimensional information projected holographically so that it appears real in a similar way to how a computer desktop interface is made up from tiny pixels, which nevertheless can display three-dimensional images. However wild this may sound, this concept is supported by many prominent and well-respected physicists.

Every theory necessitates assumptions. The aim of new theories should in part be to explain the assumptions of previous theories and so develop a more fundamental understanding of reality. For example, two of Einstein's assumptions in his theory of relativity are that objects with no mass travel at the speed of light and that every observer sees the same maximum speed of light. These were not findings that emerged from the theory of relativity, but instead formed part of its foundation. When Hoffman reported recently on progress, he described how his mathematical formulation of conscious agents can directly derive these assumptions from his mathematical expressions of conscious agents' experiences and actions.[326]

Hoffman is currently attempting to go beyond these promising early findings and ultimately to describe the principles that have the potential to explain every single feature in both the physical and experienced (mental) world. He has suggested that this highly ambitious venture may now have become relatively more straightforward owing to recent developments in the fields of high energy physics and mathematics. A major international effort involving researchers in these fields is exploring hypothetical structures that are said to lie outside of spacetime and that give rise to the universe we see.[327] These structures may be more fundamental than what we currently think of as so-called fundamental physical particles that lie within spacetime. They are known as 'positive geometries' and are grouped in categories known as 'decorated permutations'. Hoffman points out that if he can demonstrate how his theory of consciousness with its networks of conscious agents gives rise to these positive geometries and decorated permutations through his mathematical formulations, then conventional physics can take these forward into spacetime to derive a complete set of physical laws that unite the theories of relativity and quantum physics. And this will show, he says, that "we can start with a theory of consciousness and then boot up physics".[328]

Features of Consciousness

How might these approaches to consciousness address our three features of consciousness of rich personal experience, differing states of consciousness, and determining between conscious and non-conscious entities?

Rich Personal Experience

As both Lanza's biocentrist theory and the conscious agents theory of Hoffman suggest that consciousness is fundamental, this again arguably means that no further explanation for this feature of consciousness is possible or required. Consciousness is as it is, including its nature as rich experience. In addition, the personal nature of mind seems to flow logically from these theories because of the primacy of life in Lanza's theory and the centrality of conscious agents in Hoffman's theory.

States of Consciousness

It is not immediately obvious how Lanza's and Hoffman's theories might lend themselves to explanations for distinct states of consciousness. They do not seem to address this aspect. Hoffman has suggested that this feature—as with many other features of both consciousness and the physical world—should eventually be addressed through application of his mathematical schema, although this of course remains to be demonstrated.[329]

Determining Between Conscious and Non-Conscious Entities

Life may be key to these theories' explanations for determining between conscious and non-conscious entities in a similar manner to how Hennacy's and Sheldrake's field theories of consciousness dealt with this feature. Whilst it seems that these theories do not directly address this question, Lanza's theory of biocentrism seems to imply that something needs to be alive for it to be conscious and that anything not alive cannot be conscious. Likewise, it could be that

Hoffman's theory, with its focus on the importance of perception, also assumes that life is a prerequisite for consciousness.

Weird Phenomena

How might these theories suggesting that consciousness is all there is account for weird phenomena? The two theories we've looked at in this chapter suggest both that our consciousness may not end with the death of our physical body and that we have genuine free will. However, they are either antagonistic or silent when it comes to the other weird phenomena we've been looking at. This is despite their proposals that consciousness is more real than the physical world, a framework which, one might think, should be able straightforwardly to accommodate manifest weird phenomena, perhaps even as an entirely predictable feature of a universe where consciousness comes first. Let's examine in more detail what these theories have to say on these topics.

Consciousness Beyond the Body

The proponents of biocentrism argue that there can be no death of consciousness. To understand the reasons for this we need to examine a second interpretation of quantum mechanics that may be compatible with this theory.

Whilst, as we've seen, biocentrism emphasises the importance of the observer in collapsing the wave function to manifest physical reality—the Copenhagen interpretation of quantum mechanics—biocentrism suggests that a version of the interpretation known as 'many worlds' may also be true, despite how it is usually seen as a competing interpretation. The many worlds interpretation of quantum mechanics was first proposed by Hugh Everett in the 1950s and suggests that every time there is a choice to be made by a particle, say for it to have travelled along one path rather than another, a new branch of the universe springs into existence so that both paths are taken, albeit in separate worlds. This explanation can resolve many of the apparent strange aspects of quantum mechanics. For example, a superposition in the many worlds interpretation wouldn't mean that a particle could be in two places at once—it's just a particle having one location in one branch of the multiverse and another in a second branch. One seeming drawback of this many worlds interpretation is that it involves a continuous unfolding of an unimaginably vast multitude of universes at every single moment, one for every possible action that every single particle might take. Whilst most of those who support the many worlds interpretation imagine that each of these universes has a tangible reality, the view from biocentrism has it that the only universes that actually exist are those that accommodate an observer who can collapse its wave function. This drastically limits the number of possible physical universes with the vast majority of branches persisting only as unrealised statistical and theoretical potentials and so perhaps makes this interpretation a shade more palatable.

According to biocentrism, this many worlds interpretation implies that, from the perspective of ourselves as individuals, there will always be a branch of the multiverse to support our consciousness, and so we cannot die. This is because the universe in which we live requires our participation as observers. However, as our bodies inevitably age and decay then the number of branches of the multiverse that are able to support a functioning brain to host our consciousness will eventually reach zero. But, from a biocentrist perspective, our consciousness cannot cease to exist because it is fundamental to nature and the observer is always present and aware. Therefore, according to the many worlds interpretation within a biocentrism framework, "many other possible configurations are, and always will be, available to support your consciousness, including a world in which you find yourself two years old and living a slightly different life, that is, an alternative history".[330] Hence our consciousness is immortal.

Hoffman is rather more circumspect than the seemingly definite biocentrist view about the ongoing survival of consciousness. However, he does suggest that our consciousness may persist beyond the death of our bodies. This would be in the sense of a conscious agent (an individual consciousness) disengaging from their perceptual interface to reality when the body dies. He compares life to playing a virtual reality game of beach volleyball where a player might remove his

> headset and body suit. His avatar collapses onto the sand, inert and unresponsive. But he's fine. He just stepped out of the virtual-reality interface. When we die, do we simply slip out of the spacetime interface of homo sapiens? I don't know.[331]

It seems therefore that these theories might be compatible with consciousness surviving the death of the physical body and perhaps allowing for reincarnation. However, neither the authors of biocentrism nor Hoffman appears to have engaged with the extensive literature regarding the specifics of the relevant evidence for survival after death, some of which was summarised in Chapter 3. Therefore, arguments from these theories for the continuation of consciousness beyond bodily death could be seen as somewhat abstract and lacking any grounding in the actual phenomena that has been so compellingly documented by researchers.

Consciousness Transcending Space, Time and Matter

Both theories we've looked at in this chapter propose that our consciousness creates the physical universe, either through our observations of the world or from conscious agents. In the case of biocentrism, this is not just reality as it is in the present but also a complete and coherent history that is exactly conducive to us being here. These theories argue that modern physics is showing us that there is no such thing as space and time outside of our perception, that our consciousness can perhaps straddle multiple universes and may even be eternal. Perhaps, therefore, it is not too much of a stretch for these theories to allow for phenomena that transcend space and time such as telepathy and precognition, and that can manipulate matter, such as psychokinesis?

However, in the case of biocentrism, whilst the theory's authors maintain that the theory seems "very much to contradict the evidence of our everyday experience of separate consciousnesses",[332] in echoes of some of the critics of parapsychology, they summarily dismiss the possibility that any of these weird phenomena might be real as so much "popular nonsense".[333] They assert that thoughts cannot be shared and state that

> no one can consciously control external physical events that affect them—meaning those that inherently lie beyond human volition such as a boulder rolling down a hill into the path of their car—any more than they control whether the next coin flip will come up tails or heads.[334]

Some would say that this confident assertion against telepathy, precognition and psychokinesis is contradicted by the large extent of well-controlled experimental evidence demonstrating their reality. As for Hoffman, whilst not explicitly rejecting these phenomena, he appears to be silent on addressing the potential for such experiences to provide us with important clues about the nature of consciousness.

It seems that the evidence regarding weird phenomena has not been seriously considered in the context of these theories.

Free Will

The perspective that consciousness is fundamental seems consistent with a view that we have free will. After all, if we create the universe, either as observers or conscious agents, then this suggests that our consciousness has primacy over the physical world, that our actions and decisions in the world are purposeful and meaningful. We need not be constrained by the assumptions held by brain-based theories that physical processes determine our consciousness rather than the other way around. And the view of both biocentrism and Hoffman is that we do indeed have free will.

As we discussed in Chapter 6, findings from experiments performed by Benjamin Libet and colleagues are often cited as the most powerful evidence against free will. Libet measured what was going on in people's brains leading up to the moment when they chose to move their arm and found that neurons related to this action fire in the brain several seconds before a person becomes aware of their decision to move that limb. In other words, a physical process in the brain over which we seemingly have no control significantly precedes what feels like our free will to move. Libet concluded that our perceptions that we are able to make freely-arrived-at decisions and choices is therefore wrong because it is not borne out by what we see happening in the brain. Whilst some scientists have highlighted issues with how Libet's experiment has been interpreted (as we also saw), according to the theories discussed in this chapter, even if Libet's results are taken at face value they would not necessarily rule out us having free will.

The argument from biocentrism relates to how our consciousness creates the entirety of the universe. This is not just what exists in the here and now, but also the entire history of everything leading up to this point. The observed past is necessarily consistent with the observed present. As we saw, this can explain the exquisitely finely-tuned physical parameters that enable us to exist but it can also account for the presence of brain processes observed taking place in the moments before we feel ourselves making a choice, as in Libet's experiments. Biocentrism suggests that when we make a decision to move an arm, say, this has the effect of collapsing the wave function of possibilities so that everything we observe is consistent with that decision, even if it has already happened in the so-called past in a conventional view of time. Therefore, whilst this may contravene our commonsense notions of cause and effect, Libet's findings do not rule out free will.

Hoffman takes a more fundamental approach to Libet's experiments. He maintains that spacetime—our physical world—does not exist unless and until it is perceived by consciousness. Hence the brain and its neurons cannot be the cause of our actions because they are not real as they have not been directly observed. Furthermore, Hoffman suggests that the mathematical basis of his theory seems to be showing that free will is an integral feature of the conscious agents that give rise to our world. He suggests that the probabilistic aspects of quantum mechanics are not some objective randomness as is assumed by those who support brain-based theories of consciousness, these probabilities are really the manifestations of choices and decisions made by conscious agents. Therefore, unlike brain processes, free will is a fundamental aspect of the universe.[335]

***

This brings us to the end of our discussion of ten scientific theories where consciousness could extend beyond the brain, adding to the four theories we discussed in Chapter 2 suggesting that consciousness is a product of brain processes. In the next chapter, we'll take a breath to review all of these approaches in the round. How effective have they been in accounting for defining features of consciousness? How well could they explain weird phenomena? In addition, we'll look at whether they even qualify as good scientific theories, for example in providing a framework for testing their predictions.

Chapter 12

How Effective Are These Theories?

This book examines some of the most promising theories offering a scientific explanation of consciousness. As noted in Chapter 1, these fourteen are far from the only such proposals. In 2024, Robert Lawrence Kuhn, host of the *Closer To Truth* television and YouTube channel, published a lengthy paper that was an attempt to capture and summarise what he believed to be every single current scientific theory of consciousness,[336] which at the time numbered over 200 such models.[i] The aim of this book is to offer a deeper dive into a selection of these theories and provide a flavour of some of the frameworks proposed to explain consciousness.

Whilst I've included the theories that most regularly feature in the mainstream scientific press—the four brain-based theories, the orchestrated quantum wave collapse theory by Hameroff and Penrose, and integrated information theory—the others have been chosen mainly according to my interests and knowledge of the field. I'm not suggesting that these are necessarily leading candidates for a theory of non-local consciousness and other authors may have chosen differently. Indeed, one problem with theories that are seen to lie outside of mainstream science is that there is as yet nothing approaching consensus on which may be leading or in how we might assess them. Part of the reason for writing this book is to offer one such approach. As should be obvious by now, it is also the case that the categorisations I've suggested are somewhat arbitrary. For example, many of the

i A task he himself acknowledged as being unrealistic. Indeed, three of the theories that are included in this book—those by Neppe and Close, Hennacy, and Lanza et al.—are absent from Kuhn's paper.

theories of non-local consciousness relate to quantum mechanics, not just those in Chapter 7, others reference aspects of fields or information, not just those in Chapters 9 and 10, and so on.

Effectiveness as Explanations of Consciousness

Throughout this book, I have considered how effectively the fourteen theories work as comprehensive explanations for consciousness both in terms of the three defining features of consciousness I proposed in Chapter 1 and for the different types of weird phenomena considered in Chapters 3–6. My conclusions are of course open to debate. It is entirely possible—perhaps even likely—that others will disagree with my assessments, including the authors of the theories themselves and you yourself. This might be down to an honest difference of opinion, my making unwarranted assumptions or speculations or my misunderstanding what these theories are about. Nevertheless, a consideration of these and similar questions should offer one way to pinpoint gaps in the theories and help to identify areas where progress can be made, a subject we'll look at in the next chapter.

Theories of Brain-Based Consciousness

Let's begin with how effectively theories of a brain-based consciousness can account for the three defining features of consciousness. Some of those who reject the brain-based approach have been particularly scathing about its ability to offer any persuasive account for rich personal experience. For example, Donald Hoffman, whose theory we examined in Chapter 11, has stated, "as of now, no physicalist [brain-based] theory can explain a single specific conscious experience… We have these theories of conscious experiences that cannot explain *any* conscious experiences".[337] Likewise, the philosopher Bernardo Kastrup suggests that brain-based theories seem to require a 'magical step' in failing to "explain how an otherwise unconscious substrate [i.e., the brain] can light up with consciousness" simply because it exhibits certain processes.[338] And, whilst there are suggestions from at least three of the brain-based theories for how they might address states of consciousness and how they might be used to determine whether an entity has consciousness or not, these explanations are as yet incomplete.

As has been noted, a particular weakness of theories of a brain-based consciousness is that they cannot account for the existence of weird phenomena, even in principle. If consciousness is entirely a product of the brain, then it is challenging, to say the least, to see how it could extend beyond the body, transcend space and time or directly influence matter. This may be why these phenomena are so roundly rejected by those advocating for such a perspective. Many brain-based theorists have also concluded that free will is an impossibility due to their belief that we are entirely at the mercy of ultimately deterministic physical processes.

But there are issues with a brain-based approach beyond these gaps in explanatory power. If one believes that the rich variety of life can be entirely accounted for by evolutionary theory, that complex organisms such as humans (and our brains) have emerged through a process of natural selection occurring over millions of years, then a key question is what purpose does having consciousness serve? In his book *The Selfish Gene*, the biologist Richard Dawkins describes humans as mere "survival machines, robot vehicles blindly programmed"[339] whose sole purpose is to ensure the continuation of our genetic code. It might be reasonable to ask that, if this is the case, why is there consciousness at all? What selective advantage does consciousness offer an organism over one that has no consciousness? This is known in the field as the 'zombie problem' because it is possible to conceive of designing a machine or robot that can act in all ways like we do, perhaps even think like we do as an artificial intelligence, but that does not have consciousness or a mind like we do.[340] In other words, what is the evolutionary advantage of a consciousness that has no real effect in the world? It's a question that's hard to resolve from this worldview.

Another issue is that the architecture of consciousness described by these theories seems mostly bound up with how the mammalian brain works, particularly the human brain. This largely anthropocentric perspective where consciousness is primarily associated with humans is in keeping with a long history of Western thought and is still implicitly or explicitly held by many people. This suggests that, at a push, other animals are only likely to have consciousness if they are similar to us in some respect.[341] However, there are an increasing number of studies that point to what looks a lot like consciousness in non-human animals, such as crows being trained to report what they see,[342] cuttlefish remembering and recalling specific details of past events[343] and bees seeming to engage in play activities when they roll balls 'just for fun'.[344] Research findings such as these have led scientists to convene the Cambridge Declaration on Consciousness, which stated that "the weight of evidence indicates that humans are not unique in possessing the neurological substrates that generate consciousness"[345] and the more recent New York Declaration on Animal Consciousness.[346] More controversially, it has even been suggested that plants too may have consciousness, based on observations of behaviours such as communicating with each other, recognising close relatives, making decisions, and being able to learn and remember things.[347] With this in mind, it seems unnecessarily limiting to base theories of consciousness largely on the workings of the human brain, as do brain-based theories.

Nevertheless, many scientists believe that it is only a matter of time before we resolve how the brain creates consciousness. In 1998, the neuroscientist Christof Koch was so confident that the brain-based perspective would soon explain consciousness that he made a wager with the philosopher David Chalmers. Koch staked a case of wine that within 25 years a clear signature for consciousness would be identified amidst the activity of the brain. However, in 2023, Koch had to accept defeat. He conceded that there was so far no clear explanation of consciousness in neural terms, nor indeed in any other respect, although this didn't stop him from renewing the bet with Chalmers for the next quarter century.[348]

Theories Where Consciousness Extends Beyond the Brain

Now let's turn to theories suggesting that consciousness extends beyond the brain. Many of these non-local theories can account for rich personal experience because they begin with a view that consciousness is fundamental. Because consciousness itself is the starting point for these theories, there is no need to justify its emergence from something else. Other theories we've looked at don't directly address this question, perhaps because their focus is instead on how consciousness operates, not on its essential nature.

Most non-local theories also don't address explanations for different states of consciousness. Notable exceptions are Hameroff and Penrose's theory and integrated information theory (IIT), which could be seen as adopting a similar approach to this question as brain-based theories, that is, based on the presence or absence of their suggested mechanism for consciousness. However, again like the brain-based theories, they cannot yet account for the rich diversity of states of consciousness. As to the final defining feature of discerning between conscious and non-conscious entities, only Hameroff and

Penrose, IIT and Campbell's theories explicitly offer some attempt to account for this distinction.

In general, as might be expected, these non-local theories perform much better than brain-based theories in accounting for weird phenomena, at least in principle, although many are distinctly lacking in detail regarding precise mechanisms. However, whilst some of these theories do not preclude weird phenomena, they do not necessarily actively incorporate them into their theory. Indeed, some theorists such as Tononi and Lanza go out of their way to dismiss the very possibility of weird phenomena. By way of contrast, weird phenomena feature as an integral component of some of these theories, in particular those of Carr, Hennacy and Sheldrake.

Effectiveness as Scientific Theories

How well each theory can account for features of consciousness is only one side of the story. The fourteen theories in this book tackle the question of consciousness from a scientific perspective and do not in the main claim to be philosophical theories.[ii] Whilst both science and philosophy aim to understand the nature of the universe, they each embody a distinct approach. Science employs empirical tools such as observation and evidence-gathering through the use of experiments and other means whereas philosophy rests primarily on logic, reason and deduction. This means that their respective areas of focus often diverge. However, where their interests coincide—as they do in attempting to understand consciousness—science and philosophy should be complementary. Ideally, findings and critiques from each can be used to enlighten and inform the work of the other, ultimately leading to more robust insights to uncover fundamental truths. As noted in Chapter 1, the scientific approach to understanding consciousness is a very much more recent endeavour than philosophical investigations into the nature of mind, which date back thousands of years.

In this section, we'll look at how well our theories perform as science. In order to do this, we need a clear sense of what makes for a good scientific theory. There are several criteria we can use. Firstly, a

---

ii It has to be said that making such a rigid distinction between science and philosophy is a little simplistic. In the Appendix I discuss how our fourteen scientific theories might be interpreted using traditional philosophical classifications.

theory should aim to be comprehensive. It should be able to explain as wide a range of observations as it can or be able to offer reasons why certain aspects are excluded. Secondly, a theory should be able to reduce a large set of diverse observations to a more manageable and coherent set of principles as efficiently and succinctly as possible. In other words, it should be parsimonious.[349,350] Thirdly, a good scientific theory should draw attention to neglected or unanticipated properties. It should make unexpected or surprising predictions, which, if supported by observational or experimental evidence, would provide the theory in question with added credence. Fourthly, it should make predictions that can be put to the test. Importantly, these predictions should be falsifiable in that it should be clear how any prediction could be proven wrong.[351] As has been noted by many authors—perhaps most famously in recent times Richard Dawkins in his book *The God Delusion*[352]—some metaphysical claims, such as the existence of God, are ultimately unfalsifiable.

Let's look at our fourteen theories to see how effective they are in terms of these criteria. As we've seen from the last section, our four brain-based theories are far from comprehensive. They not only fail to account for most 'normal' features of consciousness but, because they assume that consciousness arises from, or is equivalent to, physical matter, they also cannot—even in principle—accommodate weird phenomena. Conversely, they do seem to address a wide range of observations in a parsimonious fashion, at least in terms of the limited scope of the brain. However, it is debatable whether any brain-based theory makes claims that are truly unexpected, except in a narrow neuroscientific sense.

A notable strength of brain-based theories is that they strive to make clear and falsifiable predictions that can be subjected to experimental testing. This is perhaps not surprising given that these theories were developed with the express intention of explaining experimental and clinical data, for example those arising from brain imaging. The results of tests of these theories can be found in an online database of experiments performed in relation to brain-based approaches to consciousness. This is known as the Consciousness Theories Studies database or ConTraSt for short.[353] Its aim is to provide an "unbiased, theory-neutral, quantitative and systematic review of empirical findings around leading theories of consciousness, providing a bird's eye view of the field and looking for potential biases in interpreting empirical findings".[354] At time of writing, the database incorporates 516 separate experiments examining a variety of aspects of the brain thought to be involved in consciousness, such as brain waves, interactions between brain regions and neural activity.

The ConTraSt database has enabled scientists to examine the effectiveness of brain-based consciousness research, not just in terms of bald experimental findings, but also interpretations made from these findings, the methodological choices of researchers, the starting viewpoint of experimenters and other variables. Unfortunately these analyses have found that most of these studies interpret their findings after the fact rather than conducting up-front tests, meaning that predictions are often not in practice falsifiable. Critics have also found that there is a general tendency towards confirmation bias in interpreting findings (in other words, they find what they're looking for) and that there is a lack of rigour in how findings are related to theories.[355,356]

Hameroff and Penrose's theory of consciousness as quantum wave function collapse (Orch OR) is more promising in terms of comprehensiveness, meeting as it does many of our defining features of consciousness, including potential explanations for certain weird phenomena. The theory is also relatively parsimonious and makes unexpected and falsifiable predictions. Again, its effectiveness as good science may not be surprising given Hameroff and Penrose's strong scientific backgrounds, in particular Hameroff's role in brain science. Conversely, Stapp's quantum theory of consciousness might be considered rather less rigorous. Whilst it is rooted in scientific principles of quantum mechanics, it is neither comprehensive nor parsimonious in terms of specific features of consciousness. It also fails to make predictions that are unexpected (or perhaps even clearly established at all) or that are falsifiable.

Likewise, neither of the hyperdimensional theories of Carr or Close and Neppe fare particularly well in terms of being properly scientific. As we've seen, they fail to account for many features of consciousness and they are not parsimonious. However, with some development, it is possible that they might be capable of making surprising and unexpected predictions about the nature of consciousness. In particular, Close and Neppe's theory is underpinned by what these theorists claim to be a comprehensive mathematical model, which may lend itself to a programme of empirical research.

Neither Hennacy's nor Sheldrake's theories of consciousness can be said to be comprehensive. This is because they are purposely focused on particular aspects of consciousness—in Hennacy's case, the electromagnetic aspects of the brain and body and how this might interact with non-local consciousness, and in Sheldrake's, hypothesised morphic fields encompassing certain aspects of consciousness such as memory and telepathy. Sheldrake's theory might be seen as relatively parsimonious in that it is limited in scope to how morphic fields might be involved with manifestations of consciousness. Each theory can be said to lead to unexpected predictions. For Hennacy, this includes the idea that telepathy could work through electromagnetism, a suggestion that most parapsychologists have to date largely dismissed, and for Sheldrake the idea of morphic resonance, which predicts that telepathy might be more pronounced for those who are more emotionally bonded. Their ideas can and have been subjected to clear and falsifiable experimental tests, some of which we'll look at more closely in the next chapter.

The information-as-consciousness theories of integrated information theory and Campbell differ markedly in approach. Whilst integrated information theory is limited in the range of features of consciousness it attempts to explain, it can be said to be parsimonious, as well as both unexpected and falsifiable in terms of its predictions, at least within the limitations of its avowed scope. Again this may not be surprising as, just as with the brain-based theories, it too is being developed by neuroscientists and other researchers working in established university departments and its experimental findings and procedures are included within the ConTraSt database. Campbell's theory is not under active consideration within mainstream research institutions.[iii] Nevertheless, what is intriguing is Campbell's claim that he has subjected his theory to a multi-decade programme of experimental research. Unfortunately, he provides few details, at least in the sources I've consulted, although his tests are clearly quite different from lab-based research and instead seem to based on his (and perhaps others') explorations in consciousness through out-of-body experiences. Whilst some may dismiss such an approach as unscientific, the use of consciousness itself as an experimental tool might offer a productive way of interrogating what is after all an entirely subjective phenomenon. This could offer a means to overcome usually unacknowledged barriers that get in the way of what are supposed to be more objective approaches. We'll return to this idea in the next chapter.

---

iii This is an assumption, but, given the nature of Campbell's theory, it feels like a safe one.

Lanza and colleagues' theory of biocentrism does not seem to meet many of the criteria suggested for a truly scientific theory. Whilst the theory is firmly rooted in scientific ideas, there is only a limited attempt to address multiple features of consciousness. It also cannot be said to be parsimonious. It does make somewhat unexpected predictions but it is difficult to see how these might lead to a programme of experimental research based on truly falsifiable hypotheses. Conversely, Hoffman and colleagues are developing what they hope to be a truly comprehensive scientific and mathematical model for their theory. Their initial work, whilst still in its relatively early stages, is nevertheless highly ambitious and shows some promise in being able to address many features of consciousness—and indeed of the physical world—through well-defined mathematical descriptions of conscious agents.

***

It is worth emphasising that my assessment of theories in this chapter does not incorporate any view on the plausibility of these theories. Whilst some theories seem able to account for particular features or phenomena of consciousness and qualify as good (or at least reasonable) scientific theories, this does not mean that they are correct. Merely that—perhaps—they offer some hope of genuine progress in the field. My opinion is that this exercise shows that anything approaching a comprehensive theory of consciousness remains a distant prospect. Whilst I believe that many of the theories have merit, none is entirely satisfactory. Much more progress is needed. In the next chapter, I offer some suggestions for how this might be achieved.

Chapter 13

Towards a Better Theory of Consciousness

It seems clear that we are a long way from any comprehensive and effective scientific explanation for consciousness. Whilst brain-based theories are firmly grounded in neuroscientific data and generate copious experimental findings revealing the workings of neural mechanisms, they struggle to accommodate many of the defining features of consciousness, in particular, those phenomena hinting that consciousness may extend beyond the brain. Conversely, whilst many of the theories suggesting a fundamentally non-local consciousness can better accommodate weird phenomena, many of these models all too often feel like speculative flights of fancy with limited explanatory power. Interesting and intriguing, maybe, but somewhat ungrounded and lacking in rigour.

So what is to be done? In this chapter, I offer two suggestions. Firstly, that we should seek to integrate theories of consciousness into an overarching framework. Secondly, that we should subject theories to more rigorous testing to understand more fully particular features of consciousness, especially weird phenomena.

Integrating Theories of Consciousness

Perhaps we ask too much when we suggest that any particular theory should accommodate every feature of consciousness. Maybe it's better that each theory is viewed as a less ambitious, more partial explanation for specific aspects of consciousness.[i] Whilst the fourteen theories

---

[i] To be fair, some of the theories, such as the brain-based predictive processing theory (Chapter 2) and Rupert Sheldrake's theory of morphic resonance (Chapter 9) are explicitly presented in this way.

considered in this book are diverse in scope and focus, they need not necessarily be contradictory. In fact, with certain changes of emphasis, several of them could be seen to complement each other. Let's explore some tentative ideas of how different approaches might be integrated and thereby begin to construct a framework that is more comprehensive than any single theory considered in isolation.

For example, perhaps Tononi's model of the organised integration of information as consciousness can work alongside the orchestrated quantum wave collapse of Hameroff and Penrose's theory. It could be that these two theories are describing different aspects or levels of abstraction of the same organised process that we call consciousness. Although Hameroff and Penrose argue that orchestrated quantum wave collapse is consciousness itself, it could equally be understood as providing a location and mechanism for an extended non-local consciousness to interact with neural processes. As such, this model could also be compatible with Stapp's theory through providing a neuroscientific mechanism for a wave function collapse caused by a non-local consciousness. Stapp (after von Neumann) calls this non-physical consciousness the 'abstract ego', which incidentally may also be thought of as another term for Hoffman's conscious agents existing outside of three-dimensional spacetime.

Integrated information theory also suggests that our subjective experiences in consciousness are hyperdimensional geometric information shapes. Perhaps these mathematical thought forms might find expression in one of the theories of an extended consciousness, such as manifesting across multiple dimensions of consciousness in Close and Neppe's triadic theory or as structures within Sheldrake's morphic fields.

Despite the limitations of brain-based theories as complete explanations for consciousness, virtually no one disputes the importance of the brain to the mind, at least with respect to most of the 'non-weird' situations that we commonly experience as physical beings. This is hard to deny given consistent observations of associations between particular experiences and patterns of neural activity, the so-called neural correlates of consciousness. But the reality of weird phenomena suggests that a more circumscribed interpretation of brain-based theories is justified. Such a view would mean that these theories would no longer be presented as explanations for how consciousness is generated by the brain—because this doesn't fit the facts—but instead as more modest proposals that aim to describe how a non-local consciousness is processed when it is localised within the brain.[357]

Such a perspective could offer valuable insights arising from the work being conducted by neuroscientists and other researchers on brain-based theories of consciousness that can inform and enhance theories of non-local consciousness. For example, we know that brain activity causes a multitude of electromagnetic effects, such as those described by Hennacy. Her suggestion is that electromagnetic fields and perhaps biofields might be thought of as just two of the effects of consciousness at work in the body that could explain telepathy and subtle auras. For brain-based researchers to accept a less encompassing perspective would allow their theories to be integrated within a more complete and effective schema for an extended consciousness. Of course, this can only happen if and when a wider range of neuroscientists and psychologists adopt a more open mind towards the evidence for weird phenomena that would allow them to engage in such questions unconstrained by current taboos.

These suggestions are admittedly rather rough-and-ready and are offered as examples of how commonalities and points of connection between different theories might be accomplished. Work is of course required to generate more robust proposals that can be put to the test. But I would suggest that such an approach is necessary if we are to make meaningful progress in this field.

It's worth noting that the idea of integrating diverse theories related to consciousness is not original. Examples of similar suggestions include Michael Nahm's 'parapsychological synthesis' to integrate the various theories of psychic experiences into one overarching model,[358] and Michael Kirchhoff and Julien Kiverstein who suggest that predictive processing theory may provide clues as to how an extended consciousness expresses itself in the physical and cultural world.[359]

Testing Consciousness

Whilst integrating theories could be a useful step towards providing a more complete and satisfying understanding of consciousness, we still need to subject any overarching model, along with its constituent parts, to a more rigorous testing regime in order to make progress. Brain-based theories of consciousness are the subject of numerous well-funded programmes of research at major universities around the world. However, as we've seen, the effort to construct a comprehensive and effective scientific theory of consciousness requires that we look beyond the brain. Unfortunately, significant chunks of the material produced by those advocating for theories of non-local consciousness are devoted to making a general case against a solely brain-based view. This is often at the expense of presenting evidence in support of their own theory alongside clear and falsifiable predictions. Whilst it's understandable that many non-local consciousness theorists are focused on refuting the brain-based paradigm given the current dominance of this view of consciousness, this seems to be at the cost of developing their own theories so that they might have maximum explanatory and predictive power. In other words, such theories often fail to meet the demands of what makes for a good scientific theory, as we discovered in the last chapter.

In the previous section I suggested an indicative model for consciousness that might serve to integrate some aspects of the theories included in this book. This included a number of tentative propositions that could form the basis for more in-depth investigations by those with specialist expertise and a greater knowledge of the theories than me. If these and similar proposals are deemed worthy of pursuing, they would need to be extended and developed so that more precise hypotheses can be formulated and then subjected to falsifiable experimental tests. The aim is to lead to a more robust and all-encompassing model of consciousness.

Whilst there have been many experiments conducted to demonstrate the reality of weird phenomena, some of which we looked at in Chapters 3–6, there have been relatively few focused on uncovering the nature and specific features of a non-local consciousness. Despite the paucity of such research, there have been useful historical examples from consciousness researchers that can help to guide and inspire future work in this respect. Let's take a look at some of these.

Does the Brain Act as a Filter for Consciousness?

If consciousness is non-local and pervades the universe, such as has been suggested by a number of the theories of consciousness we've discussed, why does it usually feel like our minds are our own, separated from anyone else's? Even when phenomena such as out-of-body experiences or telepathy are reported that might indicate an extended and shared consciousness, these are for most people exceptional experiences rather than the norm. If consciousness can operate without the body, transcend space and time, and affect physical matter, sometimes in quite dramatic ways, why is it that this doesn't seem to happen routinely and why is it so difficult to measure weird phenomena in the laboratory?

As first discussed in Chapter 8, around the turn of the nineteenth and twentieth centuries, a number of prominent intellectuals including William James, the founder of modern psychology, and the philosopher Henri Bergson suggested that the reason for this might be that the brain operates as a conduit and filter for a wider consciousness that extends well beyond the brain. James, Bergson and others wondered whether the brain might suppress access to realms beyond the physical world and dampen weird phenomena because otherwise these experiences could be overwhelming and impair our ability to function in everyday activities.[360] As James wrote, "our normal waking consciousness, rational consciousness as we call it, is but one special type of consciousness, whilst all about it, parted from it by the filmiest of screens, there lie potential forms of consciousness entirely different".[361]

The neuroscientist Morris Freedman and colleagues hypothesised that, if this is the case, such a filter might be found in the frontal brain lobes. In order to test this theory, they asked two people who had damaged left frontal lobes, one by dementia and the other by a brain lesion, to attempt mentally to affect the behaviour of an arrow on a computer screen.[362] The arrow was programmed to move either to the right or the left of the screen depending on whether there were more ones or zeroes being produced by a random number generator (RNG), in a similar manner to the psychokinesis experiments described in Chapter 5. When no one was attempting to interact with this system, it produced results that matched what would be expected by chance. However, when the participants held an intention to move the arrow, they achieved a significant effect that went well beyond that normally seen in psychokinesis RNG studies involving participants without a brain impairment.[ii]

As a follow-up experiment, the team went on to use medical equipment known as repetitive transcranial magnetic stimulation (rTMS) that can be used to inhibit for a short time the left middle

---

ii The word 'significant' in the context of statistical findings means that this result is unlikely to have occurred by chance. See the footnote in Chapter 5 for a full definition.

frontal region of the brain in people who are healthy.[363] They tested just over one hundred participants and found that they too were able to produce movements in the arrow's behaviour using their mind alone that was again significantly better than expected by chance. As well as lending support to the brain-as-filter idea, this approach could also enable scientists to perform parapsychological tests on whether consciousness can influence matter—and indeed transcend space and time through telepathy and precognition—with much more consistency and reliability than has so far been the case.

Is it Possible to Measure Global Consciousness?

In Chapter 5, I described the results of the Global Consciousness Project (GCP). Conceived and established in 1997 by a team led by Roger Nelson, the GCP consists of a network of computers sited in various countries around the world that are continuously running random number generators (RNGs) aiming to detect the fluctuations of a supposed collective consciousness. Results appear to show measurable physical effects in the form of significant deviations from chance that coincide with large numbers of people being focused on noteworthy events such as major disasters or mass meditations.

However, there are some who question the hypothesis that these results are caused by humanity's collective consciousness acting on the RNGs. Peter Bancel, a former member of the GCP team, has analysed the project's data to look at global events that were not considered at the time by the GCP researchers but which should in principle have affected the data in the same way. For example, the GCP only looked at a selection, but not all, of the meditations held for the environmentally-focused Earth Day that takes place every year. It might be expected that there should have been a similar deviation from randomness in the RNG data for the events that weren't originally considered by the GCP team as for those that were. However, this was not the case. The additional Earth Day meditations that Bancel looked at seemed to show no effect. Bancel concluded that the GCP is not in fact measuring a field effect from global consciousness but is instead showing micro-psychokinetic effects on the RNGs caused directly by the intentions of Nelson and the other GCP experimenters. In other words, he suggested that the project is essentially repeating findings from research showing that individuals can affect RNG outputs.[364]

Richard Broughton has suggested a way to resolve the question of whether RNG devices are measuring the effects from a collective field or from individuals. This is for the GCP devices to create two random data streams at the same time.[365,366] One of these would be analysed by the GCP experimenters according to current practice to see if there are shifts in the data. The other simultaneous data stream would be hidden from these experimenters and would be analysed instead by those who are sceptical that the GCP is measuring collective consciousness. If both sets of results show the same correspondence between events receiving mass attention and shifts in randomness, then this would support the collective consciousness hypothesis. If only the set of data analysed by the GCP experimenters shows a deviation from chance, then this would support Bancel's conclusions and show that the GCP is not a collective consciousness detector.

The GCP has recently launched a new iteration of the project, known as GCP 2.0. This involves a much larger number of computers and RNGs than the original GCP project as well as implementing more up-to-date technologies and standards to ensure true randomness.[367] In order to see whether GCP 2.0 might be planning to put Bancel's concerns to the test, I reached out to one of the members of its research team, Dean Radin. He informed me that the design of GCP 2.0 has "from the start… implemented ways to test the influence vs. selection interpretations, but to avoid contaminating that test, the precise methods will be revealed at a future time".[368]

Does Consciousness Collapse the Quantum Wave Function?

The biologist Stuart Kauffman and parapsychologist Dean Radin have together formulated a model for consciousness that is fundamentally quantum mechanical.[369] Their suggestion is similar to Stapp's theory of consciousness that we met in Chapter 7 in suggesting that non-local consciousness is responsible for collapsing the quantum wave function and is what gives rise to our physical reality. As we saw, this is based on the Copenhagen interpretation of quantum mechanics.

Radin and other colleagues have conducted experiments to test whether this supposed role of a conscious observer might be detectable within a laboratory setting. These experiments involve a modification of the famous double-slit experiment, which was also described in Chapter 7. In Radin's set-up, participants were asked to direct their attention to the double-slit apparatus and attempt to influence the path of individual particles using mental intention alone. The six initial experiments testing this consciousness-causes-collapse hypothesis showed a combined highly significant effect in the predicted direction, suggesting that this finding was very unlikely to have occurred by chance alone.[370] Interestingly, only experienced meditators who have been engaged over many years in a consistent practice to focus their mind were able to achieve this effect. Non-meditators didn't manage to achieve results that differed from chance.

A total of 28 experiments with a similar set-up have since been conducted. Eleven out of these have also reported significant outcomes, including by three groups independent of Radin's, where just one significant result would be expected by chance.[371] Radin and his co-authors concluded that "the results appear to be consistent with a consciousness-related [Copenhagen] interpretation of the quantum measurement problem".[372]

The novelty of this set of experiments is in its modification of a standard experiment used to demonstrate quantum effects, which may help to make its findings more appealing to mainstream science. However, before we conclude definitively that this means that consciousness collapses the wave function, there are other possible explanations. It could be that these results simply provide further evidence of psychokinetic effects to add to those discussed in Chapter 5, except at the level of individual particles. It is also possible to explain these results using other interpretations of quantum mechanics where there is no pivotal role for an observer, for example where consciousness influences what ultimately becomes physical reality in the transactional interpretation,[373] or where an individual consciousness influences which of many parallel universes to occupy in the many worlds interpretation.[374]

All of which suggests that more work is required to develop experiments with clearer and better demarcated hypotheses if we are to confirm experimentally the suggested central role of an observer in collapsing the wave function and thereby offer support for quantum theories of consciousness like Stapp's. However, given the persistence of a large number of interpretations of quantum mechanics, all of which are consistent with over a century of experimental observations, this is admittedly no small task.

Is Seeing the Future Best Explained by a Universal Mind or by Quantum Effects?

Our next example of an experiment to test non-local consciousness using weird phenomena also comes from Dean Radin, building on his and Kauffman's suggestion that consciousness is in some way quantum mechanical. As we saw in Chapter 4, there is a significant body of evidence suggesting that it's possible in some circumstances to foresee the future without recourse to the normal senses or by deduction. When this happens consciously, for example in a dream, it's known as precognition; when unconsciously, for example through pupil dilation or a change in heart rate, it's known as presentiment. Kauffman and Radin suggest that because quantum mechanics involves effects that are not localised in space and time, if consciousness is in some way quantum, then non-local weird phenomena such as precognition and presentiment "may be viewed as not only plausible, but as expected to occur".[375]

In particular, Radin has hypothesised that the quantum feature of entanglement may be key to how precognition and presentiment might work. His suggestion is that seeing the future may occur through the mind becoming entangled with a future version of itself through quantum effects, and it is this that is the root explanation of precognition and presentiment. This implies that the mind must be able to experience its future self in order for quantum entanglement to be present. If Radin's quantum hypothesis is correct, then it should not be possible for the mind to gain access to information about the future if it is never exposed to (entangled with) this future.

In order to test this proposition, Radin decided to include a hidden feature within a long-running public online psychic testing website run by the Institute of Noetic Sciences, known as the Psi Arcade or 'Got Psi?'.[376] Since its inception in the mid-2000s, the Psi Arcade website has hosted 85 million trials involving 250,000 people. This extremely large dataset allows for the detection of very subtle effects. The hidden feature Radin incorporated was that in some precognition tests the participant was told which picture was the target they were attempting to foresee and in other tests they weren't. Radin's expectation was that the participants who received feedback about the target should perform better than those who hadn't because their minds would be accessing future information through quantum entanglement between their brain in the present and their brain in the future.

Precognition about a future event shouldn't be possible without experiencing feedback.

Contrary to his expectations, Radin's results seem to show that feedback is not necessary to demonstrate precognition. It didn't matter whether or not participants saw the target after the test for them to perform better than expected by chance. This appears to provide evidence that some mechanism is involved in precognition other than, or in addition to, quantum entanglement effects. Radin's preliminary conclusion is that his findings may point to a model of the mind accessing universal or holographic information rather than operating entirely according to quantum mechanical principles.[377]

This is an excellent example of an experiment about non-local consciousness that has clear constraints and a falsifiable hypothesis. (Indeed, Radin's hypothesis *was* falsified.) More work is of course needed to clarify and confirm Radin's conclusions. This might include, for example, conducting experiments that involve more immersive approaches to facilitating precognition in order to generate larger effect sizes than we typically see from people being tested in remote online programs where it's impossible to know how focused participants might be on the task. One option might be to use the ganzfeld set-up as has been used in precognition experiments like those discussed in Chapter 4.

Future experiments on the nature of precognition and consciousness might also usefully focus on the so-called 'intervention paradox' that is a feature of some of these weird experiences.[378] This occurs when the person who is seeing the future takes some action to prevent what they foresaw happening, usually because of some undesirable consequence such as injury or death. We encountered one example of this in Chapter 4 when Rachel managed to prevent the precognitive nightmare of seeing her daughter hurt by shattered glass because she recognised in waking life the situation of the dream unfolding before her. The paradox of course is that intervening in this way means that the future event that was the cause of the precognition (her daughter being injured) is no longer there. So what was the event that caused the dream? Conducting this sort of experiment involving an element of choice to make an intervention could also be used to shed light on the nature of free will as well as that of precognition.

Does Psychic Information Make Use of Electromagnetic Waves?

Ever since the late nineteenth century when weird phenomena first began to be subject to rigorous scientific investigation, there have been suggestions that certain aspects of these phenomena might be related to electromagnetism.[379] In particular, it was suggested that information transmission between people (telepathy) or from remote places (clairvoyance or remote viewing) might take place via electromagnetic waves, especially after electrical brain waves were successfully detected by Hans Berger, the inventor of the electroencephalogram (EEG).[iii] As we have seen with Sheldrake's and Hennacy's theories (Chapter 9), some researchers still suggest that consciousness may be related to fields, albeit not necessarily electromagnetic.

Whilst the hypothesis associating psychic information with electromagnetic waves was seemingly disproved, in particular by Soviet researchers in the 1960s conducting successful ESP experiments with participants in Faraday cages and at great distances, both of which shield most electromagnetic waves, there remained the possibility that psychic signals might still be conveyed by Extremely Low Frequency (ELF) waves because these can pass through most physical structures and be transmitted over large distances. As discussed in Chapter 9, Hennacy has proposed that ELF waves might be the conduit for psychic signals, pointing out, in particular, that the frequency of theta brain waves, which have often been associated with psychic phenomena, make them essentially ELF waves.

In order to test the hypothesis that psychic information is conveyed by ELF waves, in the 1970s, Stephen Schwartz conducted remote viewing experiments involving participants on a submarine at a depth below the sea of 360 m and at a distance of 800 km.[380] The participants were able successfully to discern information about remote locations through psychic means. Whilst the submarine was not deep and remote enough definitively to rule out the possibility

---

iii It's interesting to note that Berger's initial interest in the topic of brain waves was prompted by his experiencing a case of apparent spontaneous telepathy. This occurred when his sister had a strong feeling that he was in danger and had her father send Berger an urgent telegram at the exact same time as he was thrown from his horse into the path of heavy machinery whilst serving in the German cavalry. (Related in İnce, R., Adanır, S.S. & Sevmez, F. (2021) The inventor of electroencephalography (EEG): Hans Berger (1873–1941), *Child's Nervous System*, 37 (9), 2723–2724. https://doi.org/10.1007/s00381-020-04564-z)

that psychic information uses ELF waves, which would require a sea depth of at least 2000 m, Schwartz nevertheless suggested that these results were "highly suggestive, if preliminary".[381] Unfortunately, there appear to have been no confirmatory studies. As Schwartz more recently reflected, "it would be lovely to replicate [this experiment] but it seems unlikely to ever happen, given the scarcity of deep ocean submersibles, and their cost of operation. One thing is certain: no ELF communications expert could ever explain what happened".[382]

Whilst Schwartz appears certain that ELF waves cannot be responsible for psychic information transfer, Hennacy's theory suggests that more research may be required on this point. As noted in Chapter 9, ELF waves are extremely limited in their capacity to carry large amounts of information. Whilst this might suggest that ELF waves cannot be a conduit for psychic information, we are far from understanding the form in which psychic information might be conveyed. It is therefore conceivable that 'psychic formats' could embody highly efficient compression that would enable their transmission using ELF. Perhaps this could involve quantum communication, such as in a form suggested by theoretical physicist Latham Boyle in the context of potential messages that might be sent by alien civilisations,[383] and again point to consciousness being in some way quantum mechanical. This is of course highly speculative. Nevertheless, if we were able to replicate Schwartz's experiment at more significant depths of sea water, we might be able conclusively to dismiss or allow this possibility.

Can Morphic Fields Hold Collective Memories?

As we also saw in Chapter 9, Rupert Sheldrake proposes the existence of non-local morphic fields. I would suggest that perhaps the strongest feature of Sheldrake's whole body of work has been the development of clear predictions related to his theory along with suggestions for straightforward experiments to test them such as those outlined in his book *Seven Experiments that Could Change the World: A Do-It-Yourself Guide to Revolutionary Science*.[384] Sheldrake has suggested that morphic fields when related to physiology may be detectable in experiments. For example, people who have lost a limb through an accident often report that they retain the feeling of this limb as a phantom. Whilst they may no longer possess a physical arm or leg, they nevertheless can still feel the shape of this limb where it used to be.[385] Sheldrake has proposed a test whereby an amputee would 'touch' with their phantom limb another person who is not able to see whether this phantom touch is taking place, for example because they are blindfolded or are shielded behind a door. If that person is able to feel this phantom touch more than would be expected by chance guessing, this could be evidence for a detectable morphic field for the missing limb. In a small initial set of experiments conducted by Sheldrake, results were positive, although it is difficult to rule out telepathy as an alternative explanation.[386]

Sheldrake also suggests that morphic fields may govern the creation and maintenance of aspects of consciousness. He suggests that these fields affect individuals who resonate most closely with these fields. The strength of this resonance is related to how close any individual is with those who created or contributed to the morphic field; for example, the individual themselves, family members, members of that individual's community, their nation and their species. Morphic fields may exist, for example, that hold thoughts, memories and behaviours, and Sheldrake suggests that these consciousness fields may be collective and would therefore allow for weird phenomena such as telepathy.

Sheldrake also hypothesises that consciousness fields exist related to shared cultural or informational artefacts such as languages or the solutions for a daily newspaper's crossword puzzle. Sheldrake himself conducted a successful test of this hypothesis in the early 1980s. David Vernon and Chris Roe describe how Sheldrake himself conducted a successful test of this hypothesis in the early 1980s in which he used

> two pictures containing hidden images. According to Sheldrake, it should be easier for people to identify one of the hidden images once many others had seen it. This was achieved by presenting one of the images on television to approximately two million people in the UK. Subsequently Sheldrake found that the broadcast figure was more readily identified than a control figure when presented to individuals in countries outside of the UK (who presumably were not exposed to the broadcast).[387]

A series of recent experiments related to morphic fields for consciousness has been carried out to examine whether Western participants who have not previously encountered the Chinese language can learn genuine Chinese characters more easily than decoys that have been designed to appear superficially realistic. This hypothesis is based on the suggestion that genuine Chinese characters that have been in use by large numbers of people for thousands of years should be associated with well-established morphic fields and therefore more easily learned, whereas decoys created for the experiment should have very much weaker morphic fields and therefore should be more difficult to recall. Findings from initial experiments along these lines appeared to support the morphic resonance hypothesis,[388] although a subsequent more rigorous design with more realistic decoys failed to repeat these positive results.[389] A still more recent attempt to replicate the experiment also failed to establish the predicted effect.[390] These recent studies seem to be somewhat devastating for Sheldrake's theory, at least as it relates to language. However, this is a strength of his theoretical framework—that it is clearly falsifiable and so good science.

Conducting Science Beyond the Brain

In 2019, the Galileo Commission published a report by Harald Walach as part of its mission to encourage scientists to take seriously the idea that consciousness may extend beyond the brain.[391] The name of the Commission is inspired by the experience of Galileo, who wrote that his contemporaries were so wedded to the idea that the Earth was the centre of the universe that some refused to look through his telescope because it would have demolished their worldview.[392] This is a clear echo of some scientists today who are unwilling even to consider the evidence for weird phenomena, as epitomised by sceptics Reber and Alcock's refusal to examine parapsychological data because they are 'irrelevant', as recounted in Chapter 5.[393]

Walach and the Commission argue that an acceptance of the reality of expanded consciousness allows us to consider innovative techniques to supplement our scientific toolbox. They suggest that an enhanced understanding of consciousness requires a broadening of scientific methodology to include approaches that embody 'radical introspection' to investigate consciousness through acknowledgement of its essential subjective nature and so directly interrogate the 'structures of reality'.[394] Candidate techniques include the use of altered states of consciousness such as deep meditative practices and out-of-body experiences, such as employed by Campbell to develop his theory of consciousness (Chapter 10), intentional use of psychedelic drugs under controlled conditions and spontaneous weird experiences that seem to offer a glimpse into an extended consciousness, such as the synchronicities experienced by Close and Neppe during development of their theory of consciousness (Chapter 8).

It may also involve recruiting participants who Steven Braude describes as psychic 'virtuosi', those who seemingly have a particular marked propensity to access realms of a wider consciousness.[395] One such group of people is autistic children, some of whom appear to exhibit exceptionally high levels of telepathic skills, as Hennacy has observed in preliminary tests (Chapter 9).[396] Some of these children say that they are able to meet with each other in a shared mental space they call 'the hill' despite being physically separated, sometimes by hundreds of miles.[397] If such abilities to explore an extended consciousness are confirmed through more rigorous experimentation, this could provide an opportunity for a programme of research with suitable autistic children with the aim of uncovering valuable insights into the nature of consciousness not possible with neurotypical participants.

***

In this chapter, I've proposed that a more joined-up approach to theories of consciousness might help advance the field. It's unlikely that any of the more than two hundred scientific theories, each of which has been developed (at most) by a small group of scientists, will have all of the answers to this age-old question. Yet many of these theories offer interesting and valuable ideas that are worth pursuing. It's possible that combining the most promising theories into an overarching framework may help to create something that is much more than the sum of its parts.

In order to move towards this goal, a more rigorous approach is required to formulate theories, and develop and test hypotheses related to aspects of these theories, especially for theories of non-local consciousness. I've provided a few examples above. Particularly exemplary is the work of Dean Radin and Rupert Sheldrake, two researchers who have consistently proposed inventive ways in which we might test features of non-local consciousness. These scientists have designed and carried out experiments designed to unlock the nature of consciousness that go well beyond the question of whether weird phenomena are real.[iv] It would be beneficial for other theorists

[iv] This is not at all to belittle the efforts of researchers who conduct experiments designed to show the reality of weird phenomena or particular features of these that are not directly related to theories of consciousness. More robust and reliable findings of this nature are still

to adopt a similarly rigorous approach to theory formation and hypothesis development. Such work might usefully take place in the context of initiatives such as the 'hypothesis refinery' convened by Simon Duan, Vasileios Basios, David Lorimer and Mel Larrosa of the Galileo Commission, which aims to provide what they describe as a "collaborative forum where [consciousness] theory developers present their models for structured feedback and critique" from fellow scientists.[398]

***

Throughout the course of this book, we've considered a wide range of proposals regarding the nature of consciousness from those maintaining that it is the brain alone that creates consciousness, to those suggesting that consciousness might be quantum mechanical, hyper-dimensional, field-like, informational or ultimately all there is. So where does this leave us? Is this simply an interesting intellectual voyage of discovery or might this question be more than merely academic? Could the hints provided by weird phenomena—that consciousness seems to extend beyond the brain and perhaps be more fundamental than our physical reality—have tangible implications for how we live our lives, both as individuals and for humanity as a whole? This is the focus of our final chapter.

---

needed to advance our under-standing and to convince those in the scientific community who remain sceptical.

Chapter 14

What Does It All Mean?

As I speak with people about the topics in this book, I'm often asked which is my favourite theory of consciousness. As already discussed, each theory has at least some merit and integrating diverse approaches should serve to benefit the field. I also believe that the case for weird phenomena is overwhelming, which suggests that any satisfactory view of consciousness should also incorporate the idea that our minds reach beyond our skulls. However, I would go further to say that I believe that the whole of reality is essentially unified consciousness. This means not only that consciousness comes first but is perhaps all there is. Of course, this still leaves open many questions of how consciousness might function with respect to the brain and our shared reality, but I believe that this is the perspective that fits the evidence best and upon which it may be most fruitful to build.

My hunch is based on more than the compelling findings of scientific enquiry into weird phenomena, such as those summarised in Chapters 3–6. One day when I was around twenty, some friends and I ventured into the countryside in search of magic mushrooms, innocuous-looking organisms that spring up in wet fields in the autumn and contain the psychedelic compound psilocybin. Taking this substance promised us what we believed would be a mind-expanding trip, although we had little real notion of what that might mean. Later that evening, an hour or so after eating around fifteen of them (they tasted disgusting), I had one of the most amazing experiences of my life. The world seemed utterly changed. I didn't hallucinate and superficially things appeared much the same, yet everything was fundamentally different. I can't remember the feeling exactly—it was literally indescribable—but I do recall a seemingly irrefutable knowing that there was more to the universe than everyday physical reality.[i]

I've had several other experiences that suggest to me that there is more to our minds than the brain. Two that stand out are when I was seemingly 'visited' in a dream by someone who was dying. These dreams were unlike any other dream I've had before or since and each had the same distinctive quality—they seemed to take the form of a kind of two-dimensional monochrome. The first involved my grandfather who died when I was around fifteen; the second my manager when I was in my mid-twenties. Most startlingly, my manager's death occurred at an unexpected time—I knew he was ill but not that he was about to die. The dream was so portentous that I had made sure to note the night I had the dream, which I was later able to confirm was the same night he had died.

All of which is to say that not only do I think that consciousness extends beyond the brain, I believe I have had direct glimpses of this reality. In this chapter, I'll offer some personal reflections on what such an extended consciousness might mean for our future—what it feels like to those who have had vivid experiences of this, how a shared consciousness might once have been the common experience of humanity and how embracing our interconnectedness might help to address our current existential crises.

Oneness

The philosopher Bernardo Kastrup expresses with great clarity and eloquence a version of the idea of consciousness as essential oneness. Kastrup proposes that the universe is made solely of cosmic consciousness. He suggests that all natural physical phenomena including material objects, energy and fields arise from this cosmic consciousness. Our experiences as individual beings and those of other living organisms exist as distinct localised patterns within cosmic consciousness, similar to how a wave in the sea is a thing in its own right but nevertheless remains part of the whole ocean. Kastrup calls his theory Analytic Idealism.[399,400]

---

i It's important to note that several of my subsequent attempts to repeat this state were unpleasant in the extreme. On one occasion (my last) I felt like I would never again be sane. It was truly terrifying. Bottom line, I wouldn't recommend psychedelics except under highly controlled circumstances.

This perspective bears a striking resemblance to the metaphysics of some ancient spiritual philosophies originating in the East, particularly, perhaps, Advaita Vedānta, in which the whole of reality is an all-pervading consciousness and where the physical and mental worlds have no existence apart from the ultimate reality of an holistic consciousness that underpins everything.[401] In Buddhism, this idea is expressed most clearly in the Mahāyana tradition, which suggests that what we think of as the external world has no objective reality and that the essential 'illusoriness' of the world is what makes weird phenomena possible.[402] There have been some who suggest that the notion of an all-pervading consciousness is the ultimate basis for all religious and spiritual traditions, despite cosmetic differences between them in nomenclature, rituals and practices. The view that all religion shares a common underlying metaphysics is sometimes known as the 'perennial philosophy'. According to Aldous Huxley, at the heart of this perennial philosophy are four basic tenets:

> [Firstly that] the phenomenal world of matter and of individualized consciousness... is the manifestation of a Divine Ground... [Secondly that] human beings are capable not merely of knowing *about* the Divine Ground by inference; they can also realize its existence by a direct intuition... [Thirdly that] man possesses a double nature, a phenomenal ego and an eternal Self, which is... the spirit, the spark of divinity within the soul... [Fourthly that] man's life on earth has only one end and purpose: to identify himself with his eternal Self and so to come to unitive knowledge of the Divine Ground.[403]

If it is the case that our minds are essentially a unified 'Divine Ground', an extended consciousness, why is it that we don't tend to experience this unity and wholeness? If reality is ultimately a shared consciousness, why are we not able freely to explore domains beyond our bodies, be constantly aware of the thoughts of others or of what might happen in the future, or be able to control matter with thought? Through the course of this book, we've occasionally touched upon one possible explanation—namely that the brain might work to filter out this more expanded consciousness. According to this view, our brains shield us from what would otherwise be an overwhelming—and ultimately disabling—awareness of the immensity of the combined thoughts and feelings of every other being with whom we share our collective consciousness. As the French philosopher Henri Bergson put it: "between ourselves and our own consciousness a veil is interposed."[404]

But what if this veil was lifted? What would we experience then? In his book *The Leap*, Steve Taylor relates accounts from people who seemingly have had direct experience from beyond this veil.[405] One such comes from Mildred Norman, a political activist who became known as Peace Pilgrim who, whilst she had long felt a 'powerful spiritual force' within her, had the following experience whilst out walking:

> All of a sudden I felt very uplifted, more uplifted than I had ever been. I remember I knew *timelessness* and *spacelessness* and *lightness*... Every flower, every bush, every tree, seemed to wear a halo. There was a light emanation around everything and flecks of gold fell like slanted rain through the air... I knew before that all human beings are one. But now I knew also a oneness with the rest of creation... And, most wonderful of all, a oneness with that which permeates all and binds all together and gives life to all. A oneness with that which many would call God. I have never really felt separate since.[406]

Descriptions of such a unified and expanded consciousness have been reported as spiritual experiences throughout the ages. In what is sometimes said to be her greatest work, the sixteenth-century Christian mystic St Teresa of Avila describes the soul as a castle with God at its centre and comprising seven mansions representing stages of spiritual development.[407] She writes that "the gate by which to enter this castle is prayer and meditation",[408] and by such spiritual work ecstatic visions may occur:

> The splendour of Him Who is revealed in the vision resembles an infused light like that of the sun covered with a veil as transparent as a diamond... The soul to whom God grants this vision almost always falls into an ecstasy... this apparition is more lovely and delightful than anything that could be imagined even though any one should live a thousand years and spend all that time in trying to picture it, for it far surpasses our limited imagination and understanding.[409]

Some suggest that access to a wider consciousness might be achieved using psychedelic drugs.[ii] Huxley wrote about his experiments with mescaline in *The Doors of Perception*, describing how taking the substance seemed to lead to a dissolution of his ego, the disappearance of

---

ii Apart from the fact that taking drugs comes with significant risks to health, as noted earlier in this chapter, many authors suggest that the use of artificial means to attain altered states might be detrimental because they are shortcuts that can be taken without the spiritual development required fully to integrate experiences.

spatial relationships and a timeless state in which he experienced no past or future but only a continuous present.[410] In a direct reference to his belief that psychedelics may serve to disable the brain's filter to an expanded consciousness, Huxley's book takes its title from the poet and artist William Blake who wrote that, "if the doors of perception were cleansed everything would appear to man as it is, infinite".[411] David Luke has noted that psychedelics are often associated by those who use them with the "experience of becoming one with everything in the universe",[412] as well as with an increased incidence of paranormal experiences, such as when the parapsychologist Stanley Krippner apparently foresaw the assassination of President Kennedy whilst on psilocybin.[413]

The hypothesised brain-based consciousness filter may also be disrupted by some sudden and traumatic emotional event or crisis in what is known as a spiritual emergency.[414] Steve Taylor describes the experience of Marita who was experiencing post-natal depression and was also having to look after a fractious toddler. She hadn't slept for a week when her husband began raging, smashing objects and tearing at curtains on finding he had no clean clothes to wear. This brought on what she calls a 'shift' in her consciousness:

> His explosion, and the release of his anger freed something tremendous in me! It was at that moment that I had an experience of enlightenment... This experience was multidimensional... Spiritual knowledge was revealed to me... I understood about why mystics have talked about 'for all eternity'. In a mystical consciousness, within that one instant, you sense forever and ever, and are forever changed... I felt like I had broken through a barrier.[415]

Such accounts of what might be called 'spiritual awakening' are relatively unusual, although Taylor notes that when he began seeking these out for his research, he was "surprised how easy it was to find cases".[416] In the academic literature, spiritual and paranormal experiences tend to be referred to using terms such as 'anomalous' or 'exceptional' implying that these are rare, or at least not entirely commonplace. But what if this was not always so?

The Fall

Before the onset of civilisation around 8,000 years ago,[417] it is almost certain that people saw the world as infused with consciousness. The view that natural features such as mountains, rivers, plants and animals are as alive as we are is known as animism and is virtually ubiquitous among indigenous peoples living today.[418] Animism suggests that humans have an interpersonal relationship with the world, one in which boundaries between subject and object are indistinct and where the external world is not separate from the mental world. Animist cultures typically involve shamanistic practices where certain individuals claim to be able to access other worlds and perform feats we might describe as paranormal such as energy healing, control over the weather and contacting discarnate spirits.[419]

There are other striking features of pre-civilisational animist societies. Whilst the seventeenth-century philosopher Thomas Hobbes famously described life in ancient times as "nasty, brutish and short", Christopher Ryan suggests instead that

> there is every reason to conclude that human life in the Pleistocene [the period between 1.8 million years and 10,000 BCE] was—relative to our own lives and even more so when compared with the lives known to Hobbes and his own contemporaries—rather low-stress, communal, peaceful and rich in many important ways.[420]

Anthropologists have discovered that our ancestors typically managed to forage and hunt for sufficient food in less than twenty hours per week, much less than the typical working week of modern societies. The skeletons of ancient hunter-gatherers have been found to be large and robust and show relatively few signs of degenerative diseases and tooth decay, probably due to their having healthy diets consisting of a wide variety of raw and fresh fruits, nuts, vegetables and little meat. There seems to be no evidence of warfare and communal violence appears to have been almost non-existent. These were also fundamentally equal societies where power was dispersed amongst small groups and women had as much power as men.[421]

Jared Diamond argues that the end to humanity's 'golden age' came with the invention of agriculture, which, contrary to popular wisdom, served to make life much worse for the majority of people due to three main reasons. Firstly, diets became less nutritious as they were based on fewer types of typically more starchy plants than our hunter-gatherer forbears had become used to. (Most people's diets today are dominated by just three plants—wheat, rice and corn—none of which provide the complete range of nutrients sufficient for good health.) Secondly, basing one's subsistence on a limited number of crops means increased vulnerability to the weather meaning that starvation becomes more likely if these crops should fail. Thirdly, agriculture tends to mean people congregate in larger and more crowded societies and so are more susceptible to infectious diseases and parasites. Diamond suggests that archaeological discoveries show that agriculture "was in many ways a catastrophe from which we have never recovered. With agriculture came the gross social and sexual inequality, the disease and despotism, that curse our existence".[422]

Steve Taylor suggests that these detrimental changes in humanity's circumstances are the origin of folk tales of 'The Fall', typified by the biblical story of Adam and Eve being ejected from the Garden of Eden, which are widespread among many peoples:

> [These] tell us exactly the same thing as the archaeological evidence: that there was an ancient time when human beings lived in harmony with each other and with nature, when life was much easier and more pleasant, and when there was no war, and no selfishness or fear. But we fell away from this idyllic state. Human nature became corrupt, and life became full of suffering.[423]

History since then has been one of the encroachment of disharmony, inequality, violence and the oppression of fellow humans and nature. This spread across the world through the political and religious invasions and colonisation by European countries of Africa, the Americas and Australia and with it the decimation of indigenous peoples and a marginalisation of a more expanded view of consciousness expressed through animism.

It is important that we don't romanticise ancient and indigenous societies. For example, some Native American tribes have engaged in what was often intense conflict with each other, although this was likely due at least in part to the disruption and enforced migration that resulted from European settlers taking their land. Indigenous societies are also not immune to practices that are morally objectionable, such as the shocking gang rape recounted by Daniel Everett in his observations of the otherwise spiritual and largely peaceful Pirahã tribe of central Brazil.[424] Nevertheless, it seems likely that our ancestors had, and contemporary indigenous peoples continue to have, a much more intimate connection with each other, with nature and with an extended consciousness than most of us who live in the modern world.

Our Shadow

The transpersonal psychologist Michael Daniels writes that "in many … traditions, light is considered to be a symbol for consciousness. Light illuminates our world and brings it into awareness, thereby enabling us to act with intention and rational intelligence".[425] Daniels describes how the corollary to this is the shadow—the dark places within ourselves both individually and collectively which, if left unacknowledged and untended, threaten to destroy us.

Rather depressingly, it has become a cliché to note that we live in dangerous times for humanity and the other species with whom we share our planet. For the first time in human history, we have the capacity not only to end civilisation but also much of life on Earth. The primary threats we face come in the forms of huge stockpiles of nuclear weapons; environmental breakdown from climate change, pollution, biodiversity loss and mass extinction; pandemics made more likely by the intensive farming of animals[426] and the possible release from laboratories of genetically-engineered pathogens designed to be as virulent as possible; and potentially catastrophic unintended and unforeseen consequences that come from allowing artificial intelligence control over increasing aspects of our lives, particularly weaponry, regardless of whether or not it might possess consciousness.[427] Exacerbating these interrelated risks, we are experiencing social and political polarisation alongside eye-watering levels of inequalities of wealth and power.[428] This is humanity's shadow.

I believe that our existential crises are ultimately a consequence of the brain-based view of consciousness. This mostly implicit but wholly dominant narrative (at least in the West) maintains that we live alone in our brains and that our minds are locked in our skulls. The denial that we have free will ultimately leads to fatalism—that there is nothing we can do to save the world because we have no agency. If what we are is essentially just a body, then only its immediate survival is important. Material acquisition takes precedence over spiritual growth. Charles Tart calls this the Western Creed, a view that has selfishness and hedonism as its logical conclusion. Another frequent aspect of this perspective is adopting a position of 'scientism', an excessive adherence to materialist science as the only lens with which to view the world and a dogmatic rejection of any evidence suggesting we might be more than just neuronal processes.[429]

Many would have it that humans are naturally competitive and violent and that this extends to our evolutionary ancestors.[430] Whilst it occasionally serves us to help others, especially those who share our DNA or if we feel that cooperation may further our own interests, we remain in essence simply vehicles for the efficient reproduction of selfish genes.[431] Of course, not everyone who believes in a brain-based view subscribes to anything approaching such nihilism, but, I would argue, it remains difficult from this perspective to explain satisfactorily the origins of morality, altruism, compassion and empathy.

Humanity Rising

Might it yet be possible to reclaim our inheritance and resolve our current existential woes? As we have seen throughout this book, scientists are beginning to consider what relativity and quantum mechanics and the reality of weird phenomena might mean for our understanding of consciousness. I believe that these insights will eventually lead to a more widespread acceptance that we are part of a unified consciousness in which our minds are fundamentally interconnected. This could have profound implications for how we treat our fellow humans and the natural world and lead to a more inclusive and expanded view of human nature, which, as we've seen, may not always have been one of violence and strife. Instead, we might embrace a future of peace and love underpinned by an awareness that we are all one and share an extended and holistic mind.

Such a future is of course far from given. But it may be within our grasp if we embrace approaches that better align with unified consciousness. We could do worse than learn from indigenous cultures, for example, in how we speak and think. Zoë Joncheere has observed that indigenous languages tend to have fewer personal and possessive pronouns than modern languages and that these instead incorporate an emphasis on verbs, change, process and relationship.[432] She has suggested that our use of language can have a profound influence on how we view the world and proposes an alternative mode of language inspired by these ancient roots. Such an approach to language prioritises the idea of movement rather than objects so one might say and think, for example, 'walking is happening' instead of 'I am walking' or 'barking is happening' instead of 'the dog is barking'. She suggests that incorporating such a way of speaking and thinking can lead us to have "an experience of one undivided unity of interflowing movements".[133]

Indigenous perspectives are also beginning to be acknowledged as having the potential to unlock seemingly intractable problems such as tackling climate change. Land managed by indigenous peoples can deliver solutions to the climate crisis through stopping deforestation, locking up huge amounts of carbon and preserving immense biodiversity.[434] The Brazilian Minister of Indigenous Peoples, Sonia Guajajara, believes that environmental breakdown is a predictable response from the planet to a modern materialist mindset. She writes that "when humans try to dominate nature, rather than living in harmony with it, nature always responds... we must transcend our individual experiences to reach a higher state of consciousness vis-à-vis nature".[435] The experience of Wajã Xipai, an indigenous participant at a recent international meeting to drive action on climate change, speaks of how far we in the so-called developed world may need to change our current ways of working. He writes of his experience at the event:

> The voices speak of the future, of goals, of funding, coming and going with the haste of those who have a clock in their soul. In the forest, nothing is rushed: the trees have a specific time to bear fruit, there is a certain moment to plant our crops. The seeds sprout when the earth is ready. Each bird sings in its own time—there is a tinamou that sings only at dawn, at noon and at dusk, and it never comes early or late. The forest understands time as a pact between beings... Here [at the conference], time does not flourish, it is consumed. People run so much that they don't realise what is being lost between one step and another. And I think: how is it possible to save the planet without relearning the rhythm of the earth?[436]

Whilst our mindsets can learn from ancient wisdom, modern knowledge can provide us with the tools to enable a more peaceful future. Wars typically take place because of disputes over scarce resources such as oil, minerals and water. Control of land bestows immense power and wealth. Yet we now have access to incredible technological advances that could mean an end to societies based on the unsustainable extraction of resources from the land. Remarkably, we stand at the cusp of a technological revolution that could end this paradigm and offer instead a world of material abundance where there is no need to dominate or exploit others.[437]

For example, we are now able to implement energy systems, particularly solar, wind and batteries, which do not involve the ongoing extraction and burning of fossil fuels, and which also happen now to be the cheapest forms of energy in history. Such renewable energy could spell the end of dysfunctional power dynamics between countries as well as tackle air pollution and fossil fuel emissions, the main cause of climate change. As author Bill McKibben writes:

> There is a chance for a deep reordering of the earth's power systems, in every sense of the word 'power,' offering a plausible check to not only the climate crisis but to autocracy. Instead of relying on scattered deposits of fossil fuel—the control of which has largely defined geopolitics for more than a century—we are moving rapidly toward a reliance on diffuse but ubiquitous sources of supply. The sun and the wind are available everywhere, and they complement each other well; when sunlight diminishes in the northern latitudes at the approach of winter, the winds pick up. This energy is impossible to hoard and difficult to fight wars over. If you're interested in abundance, the sun beams tens of thousands of times more energy at the earth than we currently need.[438]

Other technologies mean that it is possible to produce meat, dairy and intensively-farmed crops such as vegetable oils with vastly reduced amounts of land, water and pollution.[439] Cultured meat can be grown from just a very few feeder cells. Precision fermentation can synthesise vegetable oils and make dairy proteins without the need for animals, harmful inputs such as fertilisers and pesticides, and huge extents of land (since producing food from animals globally requires an area equivalent to North and South America combined).[440] Such technologies could offer us enhanced food security, restore immense areas to wilderness and spell the end of factory farming that kills more than 100 billion animals every year having lived short and miserable lives.[441]

But can we reconcile a spiritual outlook of a unified and extended consciousness with such technological advancements? In a speech he wrote for the Nobel committee when he was awarded its prize for literature in 1927, Henri Bergson expressed his concern that the

> technological development of a society does not automatically result in the moral perfection of the men living in it... an increase in the material means at the disposal of humanity may even present dangers unless it is accompanied by a corresponding spiritual effort... antagonisms... risk being aggravated if a spiritual progress, a greater effort toward brotherhood, is not accomplished.[442]

In his final book, Bergson wrote that all great mystics

> prove to be great men of action, to the surprise of those for whom mysticism is nothing but visions, and raptures and ecstasies. That which they have allowed to flow into them is a stream flowing down

> and seeking through them to reach their fellow-men; the necessity to spread around them what they have received affects them like an onslaught of love... so that... other men will open their souls to the love of humanity.[443]

When we think of politics, we tend to think of political parties, of elections, of left and of right, and increasingly of polarisation. There is arguably a tendency for those most unsuited to exercising power to be the ones who most ardently seek it. (Hence the person best suited to rule the universe in Douglas Adams's satirical science-fiction series *The Hitchhiker's Guide to the Galaxy* is the person who least wants to do it.)[444] Many have become disillusioned with the potential for politics ever to be a force for good. Yet there have been those who secured meaningful change who have been inspired by spiritual principles. Mahatma Ghandi strived for religious tolerance and is widely credited with achieving independence for India from what was an exploitative and often brutal British colonial rule using nonviolent mass disobedience based on spiritual principles from the *Bhagavad Gita*, the foundational text for Hinduism.[445] Martin Luther King Jr., who led the civil rights movement in the United States that achieved an end to segregation based on race, expressed sentiments similar to those of Bergson. King's approach drew on principles of nonviolence inspired by his Christian faith and he said that "power without love is reckless and abusive, and love without power is sentimental and anaemic".[446]

Although it may be hostage to future events to cite any contemporary political figure, I find inspiration from the spiritual author Marianne Williamson. Williamson has twice stood to be a presidential candidate in the United States and bases her politics on the teachings of *A Course in Miracles*.[447] This remarkable spiritual work, said to have been channelled from a higher spiritual source, has a metaphysics similar to the philosophy of Advaita Vedānta in that it suggests that physical reality is essentially illusory and but a manifestation of our minds. Williamson talks of the need to address the

> dark psychic force of collectivised hatred... through harnessing the power of collectivised love... The love of which I speak is not... the woo woo, kooky thing that has been projected onto it... what we need now is a politics of love.[448]

***

Consciousness as a focus for enquiry is fundamentally different from other areas of science in that it is impossible to escape its essential subjectivity. Any attempt at approaching this topic with rigid objectivity is likely to lead to understandings that are at best incomplete and partial. All that we do, including all of our theorising and all of our research, cannot help but be conducted within consciousness. Whilst science strives for objectivity, consciousness brings us back to ourselves. Our mind is both subject and object. Regarding our collective future, this also implies that, ultimately, change must come from within. We can begin to reach towards oneness and address our shadow by recognising our common humanity and common personhood with all beings, and with the active pursuit of love, peace and harmony in all of our activities. If we truly belong to a shared consciousness, then this is surely the most valuable contribution we can make.

Acknowledgements

First and foremost, my thanks and appreciation go to Chris Roe. This book is based on an essay Chris and I wrote that won the 2024 Linda G. O'Bryant Research Prize, hosted by the Institute of Noetic Sciences (IONS).[449] I am honoured that IONS chose our essay to win, especially as the judging panel comprised many scientists who I greatly admire and respect. A number of Chris's ideas for the essay, and much of its contents, have been adopted and adapted for this book and it's all the better for it. The book—at least in its current form—wouldn't have been possible without him.

I'd like to thank Zoë Joncheere, Piero Calvi Parisetti and Claire Murphy-Morgan who read drafts of this book and provided helpful feedback. Steve Taylor, Sophie Morrison and Rosaleen McDonnell offered invaluable comments, as did Alex Thomson, who has also been on hand to offer sage and indispensable advice at all stages of what has become my academic career.

Thank you to Dean Radin for writing such a generous foreword. I'd also like to thank Arnaud Delorme of IONS for his feedback on the essay and Michael Nahm, a fellow Linda G. O'Bryant Research Prize winner who provided moral support during the intense process and was the original source for the idea of integrating diverse theoretical approaches. Thanks also to Fil Kinnersley who made an important suggestion for the final chapter.

As well as being my co-author for the essay, Chris Roe was an inspirational PhD supervisor for me and I continue to value his wise counsel, along with that of my other supervisor, Callum Cooper. It's an honour to belong to the one of the world's leading parapsychology research centres—the Extraordinary Experiences and Consciousness Studies group at the University of Northampton—of which Chris and Cal are leading members and which is populated by fabulous, talented and supportive colleagues with all their fascinating research interests.

I owe my friend Nick Tebble a huge amount for our stimulating discussions over many years about all things consciousness-related as well as his insightful comments on the essay, which prompted useful late amendments to the book. I'd also like to thank our local consciousness discussion group, Walter Gregory, Helen Steward, Rob Knipe and Mike Pilling, for conversations in the pub on the nature of mind, particularly Walter, Helen and Rob for helpful specific feedback at various points in the process.

It goes without saying that none of the aforementioned people bear responsibility for any potential errors, misunderstandings or misguided speculation that may have found their way into these pages. They're all down to me.

I'm immensely grateful to my brother Norman for introducing me to alternative views of consciousness through meditation and Indian philosophy. Last and most definitely not least, my thanks go to Rosaleen for her words of wisdom about the book (and much beyond) and of course for her amazing support, both through the process of writing the book and in life in general.

Appendix

Philosophical Classifications for Theories of Consciousness

This is a book specifically about scientific theories of consciousness. As noted in Chapter 1, religious and philosophical perspectives on consciousness have been with us for much longer than the few decades during which scientists have attempted to deploy empiricism to tackle the problem. In traditional academic philosophy, theories of consciousness are usually considered as occupying five broad categories: physicalism (or materialism), dualism, dual-aspect monism (sometimes known as panpsychism), idealism and mysterianism.[450] Each of these has an extended lineage with famous philosophical and scientific figures among their advocates. It is possible to consider the scientific theories covered in this book as subscribing to one or more of these categories.

Physicalism or materialism assumes that the physical world is all there is and that everything in the world, including consciousness, is ultimately subject to observation and measurement through the use of our senses and physical instruments. All of the brain-based theories fall into this category. Integrated information theory can be viewed as a physicalist theory and is sometimes assumed to be so. However, the most recent iterations of these ideas do not specify any specific substrate and allow for integrated information as consciousness to be expressed in a non-physical form.

Dualism is the suggestion that there are two fundamentals to the universe—physical and mental. This viewpoint suggests either that there are two different types of substances, an idea most famously defended by the seventeenth-century philosopher René Descartes, or two different physical and mental properties of a single fundamental substance of the universe. Stapp's theory of quantum effects might be considered as dualist, with his concept of the abstract ego being essentially made of different 'stuff' than the physical world.

Dual-aspect monism proposes that the universe is fundamentally only one type of substance but that this substance has both physical and mental aspects. Dual-aspect monism is also known as panpsychism where 'pan' means everything and 'psychism' means consciousness. In other words, consciousness, or at least some nascent form of consciousness, is everywhere and is in everything. This is perhaps the most 'popular' of the categories for the theories contained here. As well as Stapp's theory, which could equally be considered to be panpsychist as dualist, each of the theories of Hameroff and Penrose (wave function collapse in microtubules), Carr (hyperspace and the specious present), Close and Neppe (trinity of space, time and consciousness), Hennacy (electromagnetism and biofields), Sheldrake (morphic fields), and Tononi et al.'s integrated information theory is closely aligned to a panpsychist view.

Idealism is the view that consciousness is all there is and that the physical world only exists as a manifestation of consciousness. It is so-called because of the suggestion that the world is formed of mental 'ideas'. The physical world can be thought of either as an illusion or as something that emerges from consciousness. In viewing the entire universe as essentially made of consciousness itself or only existing because of consciousness, each of the theories of Campbell (digital information technology and evolution), Lanza et al. (conscious life creating the universe) and Hoffman (conscious agents outside space-time) is idealist.

Mysterianism is a view that rejects all of these possibilities and suggests that it is not possible to explain consciousness because humans lack the capability to do so, perhaps because any explanation of consciousness ultimately comes from within consciousness itself meaning that it is ultimately inexplicable and therefore mysterious.

Endnotes

1 *The Science of Consciousness Conference* (2025) [Online], https://consciousness.arizona.edu/science-consciousness-conference, para. 6.

2 Ibid.

3 Klostermaier, K.K. (2007) *A Survey of Hinduism*, State University of New York Press.

4 Doniger, W. (2024) Veda, *Encyclopaedia Britannica*, [Online], https://www.britannica.com/topic/Veda

5 Littlejohn, R. (n.d.) Chinese philosophy: Overview of history, *The Internet Encyclopaedia of Philosophy*, [Online], https://iep.utm.edu/chinese-philosophy-overview-of-history/ [13 June 2024].

6 Watson, R.A. (2024), René Descartes, *Encyclopaedia Britannica*, [Online], https://www.britannica.com/biography/Rene-Descartes

7 The biggest questions in science (2018) *Nature*, [Online], https://www.nature.com/collections/mnwshvsswk

8 Kuhn, R.L. (2024) A landscape of consciousness: Toward a taxonomy of explanations and implications, *Progress in Biophysics and Molecular Biology*, 190 (June 2023), 28–169.

9 Blackmore, S. & Troscianko, E.T. (2018) *Consciousness: An Introduction*, Taylor & Francis, 1.

10 Frith, C.D. & Rees, G. (2017) A brief history of the scientific approach to the study of consciousness, *The Blackwell Companion to Consciousness*, 1–16.

11 Nagel, T. (1974) What is it like to be a bat? *Philosophical Review*, 83, 435–450.

12 Tye, M. (2021) Qualia, *The Stanford Encyclopedia of Philosophy*, [Online], https://plato.stanford.edu/archives/fall2021/entries/qualia

13 Chalmers, D.J. (1995) Facing up to the problem of consciousness, *Journal of Consciousness Studies*, 2 (3), 200–219.

14 Jackson, F. (1982) Epiphenomenal qualia, *The Philosophical Quarterly*, 32 (127), 127–136. https://doi.org/10.2307/2960077, p. 127.

15 Russell, B. (1911) Knowledge by acquaintance and knowledge by description, *Proceedings of the Aristotelian Society*, 11, 108–128.

16 Block, N. (1995) On a confusion about a function of consciousness, *Behavioral and Brain Sciences*, 18 (2), 227–247. https://doi.org/10.1017/S0140525X00038188

17 Tassi, P. & Muzet, A. (2001) Defining the states of consciousness, *Neuroscience and Behavioural Reviews*, 25, 175–191.

18 Frith & Rees, 2017.

19 Hofstadter, D. & Dennett, D. (2001) *The Mind's I: Fantasies and Reflections on Self and Soul*, Basic Books.

[20] Vallance, C. (2022, June 13) Google engineer says Lamda AI system may have its own feelings, *BBC News*, https://www.bbc.co.uk/news/technology-61784011

[21] Wertheimer, T. (2022, July 23) Blake Lemoine: Google fires engineer who said AI tech has feelings, *BBC News*, https://www.bbc.co.uk/news/technology-62275326

[22] Schwartz, E.H. (2025, April 22) AI took a huge leap in IQ, and now a quarter of Gen Z thinks AI is conscious, *TechRadar*, https://www.techradar.com/computing/artificial-intelligence/ai-took-a-huge-leap-in-iq-and-now-a-quarter-of-gen-z-thinks-ai-is-conscious

[23] Vallance, 2022.

[24] Crick, F. (1994) *The Astonishing Hypothesis: The Scientific Search for the Soul*, Simon & Schuster.

[25] Seth, A.K. & Bayne, T. (2022) Theories of consciousness, *Nature Reviews Neuroscience*, 23 (7), 439–452. https://doi.org/10.1038/s41583-022-00587-4

[26] Mudrik, L., Boly, M., Dehaene, S., Fleming, S.M., Lamme, V., Seth, A. & Melloni, L. (2025) Unpacking the complexities of consciousness: Theories and reflections, *Neuroscience & Biobehavioral Reviews*, 170, 106053. https://doi.org/10.1016/j.neubiorev.2025.106053

[27] Carruthers, P. (2017) Higher-order theories of consciousness, in S. Schneider & M. Velmans (eds.) *The Blackwell Companion to Consciousness* (2nd ed.), Wiley-Blackwell.

[28] Carruthers, P. & Gennaro, R.J. (2023) Higher-order theories of consciousness, *The Stanford Encyclopedia of Philosophy*, https://plato.stanford.edu/archives/fall2023/entries/consciousness-higher

[29] Milner, D. & Goodale, M. (1995) *The Visual Brain in Action*, Oxford University Press.

[30] Mudrik et al., 2025.

[31] Baars, B.J. (2017) The global workspace theory of consciousness: Predictions and results, in *The Blackwell Companion to Consciousness* (2nd ed., pp. 229–242), Wiley Blackwell.

[32] Mashour, G.A. (2013) Cognitive unbinding: A neuroscientific paradigm of general anesthesia and related states of unconsciousness, *Neuroscience & Biobehavioral Reviews*, 37. https://doi.org/doi:10.1016/j.neubiorev.2013.09.009

[33] Yaron, I., Melloni, L., Pitts, M. & Mudrik, L. (2022) The ConTraSt database for analysing and comparing empirical studies of consciousness theories, *Nature Human Behaviour*, 6 (4), 593–604. https://doi.org/10.1038/s41562-021-01284-5

[34] Baars, 2017

[35] Mudrik et al., 2025.

[36] Seth & Bayne, 2022.

[37] Seth, A.K. (2019) From unconscious inference to the beholder's share: Predictive perception and human experience, *European Review*, 27 (3), 378–410. https://doi.org/DOI: 10.1017/S1062798719000061

[38] National Institute of Neurological Disorders and Stroke (2024) *Brain Basics: The Life and Death of a Neuron*, https://www.ninds.nih.gov/health-

---

information/public-education/brain-basics/brain-basics-life-and-death-neuron

[39] Clark, A., Friston, K. & Wilkinson, S. (2019) Bayesing qualia: Consciousness as inference, not raw datum, *Journal of Consciousness Studies*, 26 (9–10), 19–33.

[40] Kukkonen, K. (2024) Designs on consciousness: Literature and predictive processing, *Philosophical Transactions of the Royal Society B: Biological Sciences*, 379 (1895), 20220423. https://doi.org/10.1098/rstb.2022.0423

[41] Crick, 1994.

[42] Carruthers & Gennaro, 2023.

[43] Smith, M. (2020) The issues with higher order theories of consciousness, *SelfAwarePatterns*, https://selfawarepatterns.com/2020/01/02/the-issues-with-higher-order-theories-of-consciousness/

[44] Seth & Bayne, 2022.

[45] LeDoux, J.E. & Brown, R. (2017) A higher-order theory of emotional consciousness, *Proceedings of the National Academy of Sciences*, 114 (10), E2016–E2025. https://doi.org/10.1073/pnas.1619316114

[46] Mashour, G.A., Roelfsema, P., Changeux, J.-P. & Dehaene, S. (2020) Conscious processing and the global neuronal workspace hypothesis, *Neuron*, 105 (5), 776–798. https://doi.org/10.1016/j.neuron.2020.01.026

[47] Lamme, V.A.F. (2018) Challenges for theories of consciousness: Seeing or knowing, the missing ingredient and how to deal with panpsychism, *Philosophical Transactions of the Royal Society B: Biological Sciences*, 373 (1755), 20170344. https://doi.org/10.1098/rstb.2017.0344

[48] Ficco, L., Mancuso, L., Manuello, J., Teneggi, A., Liloia, D., Duca, S., Costa, T., Kovacs, G.Z. & Cauda, F. (2021) Disentangling predictive processing in the brain: A meta-analytic study in favour of a predictive network, *Scientific Reports*, 11 (1), 16258. https://doi.org/10.1038/s41598-021-95603-5

[49] Pezzulo, G., Parr, T. & Friston, K. (2021) The evolution of brain architectures for predictive coding and active inference, *Philosophical Transactions of the Royal Society B: Biological Sciences*, 377 (1844), 20200531. https://doi.org/10.1098/rstb.2020.0531

[50] Seth & Bayne, 2022.

[51] Seth & Bayne, 2022.

[52] Carruthers & Gennaro, 2023.

[53] Birch, J. (2020) Global workspace theory and animal consciousness, *Philosophical Topics*, 48 (1), 21–37. https://doi.org/10.5840/philtopics20204812

[54] Birch, J. (2022) The search for invertebrate consciousness, *Noûs*, 56 (1), 133–153. https://doi.org/10.1111/nous.12351

[55] Zacks, O. & Jablonka, E. (2023) The evolutionary origins of the Global Neuronal Workspace in vertebrates, *Neuroscience of Consciousness*, 2023 (1), niad020. https://doi.org/10.1093/nc/niad020

[56] Pezzulo et al., 2021.

[57] Lamme, 2018.

[58] Dagnall, N., Drinkwater, K., Parker, A. & Clough, P. (2016) Paranormal experience, belief in the paranormal and anomalous beliefs, *Paranthropology:*

Journal of Anthropological Approaches to the Paranormal, 7 (1), 4–14. https://doi.org/10.1163/9781848884397_002

[59] Hart, G. (2021) *Beyond Dreaming – A Guide on How to Astral Project & Have Out of Body Experiences*, pp. 19–21.

[60] French, C. (2024) *The Science of Weird Shit: Why Our Minds Conjure the Paranormal*, The MIT Press, p. 212.

[61] French, 2024, p. 214.

[62] Roe, C.A. (2025) Sceptical of scepticism, *Paradigm Explorer*, 2025 (2), 51–52. Published by the Scientific and Medical Network (https://scientificandmedical.net).

[63] Monroe, R.A. (1972) *Journeys Out of the Body*, Souvenir Press Ltd.

[64] Monroe, 1972, pp. 48–50.

[65] Nicholls G. (2025) *Navigating the Out-of-Body Experience*, The Out-of-Body & Near Death Experience Society (UK).

[66] Monroe Institute: https://www.monroeinstitute.org/

[67] Clark, K. (1984) Clinical interventions with near-death experiencers, in *The Near-Death Experience: Problems, Prospects, Perspectives*, Charles C Thomas Publishers, cited in Lawrence, M. (1997) *In a World of Their Own: Experiencing Unconsciousness*, Praeger, p. 170.

[68] Cook, E.W., Greyson, B. & Stevenson, I. (1998) Do any near-death experiences provide evidence for the survival of human personality after death? Relevant features and illustrative case reports, *Journal of Scientific Exploration*, 12 (3), 377–406, pp. 385–386.

[69] Lawrence, 1997.

[70] Willin, M. (2016) Past life regression, *Psi Encyclopedia*, https://psi-encyclopedia.spr.ac.uk/articles/past-life-regression

[71] Leo, D.G., Bruno, D. & Proietti, R. (2025) Remembering what did not happen: The role of hypnosis in memory recall and false memories formation, *Frontiers in Psychology*, 16, 1433762. https://doi.org/10.3389/fpsyg.2025.1433762

[72] Willin, 2016.

[73] Stevenson, I. (2001) *Children Who Remember Previous Lives: A Question of Reincarnation*, McFarland & Company Incorporated.

[74] Wehrstein, K.M. (2017) Pollock twins (reincarnation case), *Psi Encyclopedia*, https://psi-encyclopedia.spr.ac.uk/articles/pollock-twins-reincarnation-case

[75] Nahm, M. (2023) Climbing Mount Evidence: A strategic assessment of the best available evidence for the survival of human consciousness after permanent bodily death, in *Proof of Survival of Consciousness Beyond Permanent Bodily Death* (Vol. 5), Bigelow Institute for Consciousness Studies.

[76] Matlock, J.G. (2021) Ryan Hammons (reincarnation case), *Psi Encyclopedia*, https://psi-encyclopedia.spr.ac.uk/articles/ryan-hammons-reincarnation-case

[77] Roe, C.A. (2023) *A Representative Sample Survey of Paranormal Beliefs and Experiences*, abstracts of presented papers, 47–50.

[78] Rhine, J.B. (1926, August 9) Rhine to Saposnekov, *Parapsychology Laboratory Records 1893–1984*, David M. Rubenstein Rare Book & Manuscript Library,

Duke University, Durham, NC, quoted in Rhine Feather, S. & Ensrud, B. (2018) JB Rhine, *Psi Encyclopedia*, https://psi-encyclopedia.spr.ac.uk/articles/jb-rhine

79 Horowitz, M. (2025, May 11) The unseen legacy of parapsychologist J.B. Rhine [Substack newsletter], *Mystery Achievement*, https://mitchhorowitz.substack.com/p/the-unseen-legacy-of-parapsychologist

80 Rhine, J.B. (1937) Some basic experiments in extra-sensory perception: A background, *Journal of Parapsychology*, 1, 70–80. https://doi.org/10.30891/jopar.2018S.01.04

81 Turing, A.M. (1950) Computing machinery and intelligence, *Mind*, 59 (236), 433–460.

82 Turing, 1950, p. 453.

83 Carroll, R.T. (2015) Zener ESP cards, *The Skeptic's Dictionary*, https://www.skepdic.com/zener.html

84 Hines, T. (2003) *Pseudoscience and the Paranormal*, Globe Pequot Publishing.

85 Smith, J.C. (2009) *Pseudoscience and Extraordinary Claims of the Paranormal: A Critical Thinker's Toolkit*, Wiley.

86 French, 2024, p. 220.

87 French, 2024, p. 224.

88 Sheldrake, R. & Smart, P. (2003) Experimental tests for telephone telepathy, *Journal of the Society for Psychical Research*, 67, 184–199.

89 Sheldrake, R., Stedall, T. & Tressoldi, P. (2025) Telecommunication telepathy: A meta-analysis, *Journal of Anomalous Experience and Cognition*, 5 (1), 47–69. https://doi.org/10.31156/jaex.25934

90 Sheldrake, R. (1999) *Dogs that Know When Their Owners are Coming Home*, Crown Pub.

91 Sheldrake, R. & Smart, P. (2000) A Dog that seems to know when his owner is coming home: Videotaped experiments and observations, *Journal of Scientific Exploration*, 14 (2), 233–255.

92 Twain, M. Autobiography of Mark Twain, *Mark Twain Project Online*, https://www.marktwainproject.org/writings/autobiography3/autobiography/autobio_dictations/autobio1_1906-01-13/, pp. 275–276. This anecdote also featured in Kripal, J.J. (2020) *The Flip*, Penguin.

93 Cheung, T. & Mossbridge, J. (2018) *The Premonition Code: The Science of Precognition*, Watkins Publishing, pp. 33–34.

94 Roe, 2023.

95 Mossbridge, J., Tressoldi, P. & Utts, J. (2012) Predictive physiological anticipation preceding seemingly unpredictable stimuli: A meta-analysis, *Frontiers in Psychology*, 3 (OCT), 1–18. https://doi.org/10.3389/fpsyg.2012.00390

96 Bem, D.J. (2011) Feeling the future: Experimental evidence for anomalous retroactive influences on cognition and affect, *Journal of Personality and Social Psychology*, 100 (3), 407–425. https://doi.org/10.1037/a0021524

97 Ritchie, S., Wiseman, R. & French, C. (2012) Replication, replication, replication, *The Psychologist Magazine*, https://www.bps.org.uk/psychologist/replication-replication-replication

[98] Ioannidis, J.P.A. (2005) Why most published research findings are false, *PLoS Medicine*, 2 (8), e124. https://doi.org/10.1371/journal.pmed.0020124
[99] Engber, D. (2017, June 7) Daryl Bem poved ESP is real: Which means science is broken, *Slate*, https://slate.com/health-and-science/2017/06/daryl-bem-proved-esp-is-real-showed-science-is-broken.html
[100] Open Science Collaboration (2015) Estimating the reproducibility of psychological science, *Science*, 349 (6251), aac4716. https://doi.org/10.1126/science.aac4716
[101] Haelle, T. (2021, December 7) Dozens of major cancer studies can't be replicated, *Science News*, https://www.sciencenews.org/article/cancer-biology-studies-research-replication-reproducibility
[102] Psychology Today (n.d.) Replication crisis, *Psychology Today Blog*, https://www.psychologytoday.com/gb/basics/replication-crisis
[103] For example, in Chambers, C.D. & Tzavella, L. (2022) The past, present and future of Registered Reports, *Nature Human Behaviour*, 6 (1), 29–42. https://doi.org/10.1038/s41562-021-01193-7
[104] Wiseman, R. & Watt, C. (2025) Missing a trick: How psychology could benefit more from the paranormal, *The Psychologist Magazine*, https://www.bps.org.uk/psychologist/missing-trick-how-psychology-could-benefit-more-paranormal
[105] Hacking, I. (1988) Telepathy: Origins of randomization in experimental design, *Isis*, 79 (3,), 427–451.
[106] Bem, D., Tressoldi, P., Rabeyron, T. & Michael, D. (2016) Feeling the future: A meta-analysis of 90 experiments on the anomalous anticipation of random future events, *F1000Research*, 4. https://doi.org/10.12688/f1000research.7177.2
[107] Kekecs, Z., Palfi, B., Szaszi, B., Szecsi, P., Zrubka, M., Kovacs, M., Bakos, B.E., Cousineau, D., Tressoldi, P., Schmidt, K., Grassi, M., Evans, T.R., Yamada, Y., Miller, J.K., Liu, H., Yonemitsu, F., Dubrov, D., Röer, J.P., Becker, M., … Aczel, B. (2023) Raising the value of research studies in psychological science by increasing the credibility of research reports: The transparent Psi project, *Royal Society Open Science*, 10 (2), 191375. https://doi.org/10.1098/rsos.191375
[108] Tressoldi, P.E. & Storm, L. (2023) Stage 2 Registered Report: Anomalous perception in a Ganzfeld condition—A meta-analysis of more than 40 years investigation [version 2; peer review: Awaiting peer review]. *F1000Research*, 10 (234). https://doi.org/10.12688/f1000research.51746.2
[109] Watt, C., Booysen, T. & Pooley, A. (2025) Two ganzfeld precognition experiments with selected participants, *48th International Annual SPR Conference*, Abingdon, Oxford, UK.
[110] Targ, R. & Puthoff, H. (1974) Information transmission under conditions of sensory shielding, *Nature*, 251 (5476), 602–607. https://doi.org/10.1038/251602a0
[111] Hasted, J.B., Bohm, D., Bastin, E.W. & O'Regan, B. (1976) Experiments on psychokinetic phenomena, in C. Panati (ed.) *The Geller Papers: Scientific Observations on the Powers of Uri Geller*, Houghton Mifflin Co.

[112] Randi, J. (2015) About James Randi, *James Randi Website*, https://web.randi.org/about-james-randi.html
[113] Randi, J. (1982) *The Truth about Uri Geller*, The James Randi Educational Foundation.
[114] Wiseman, R. (2011) *Paranormality: Why We See What Isn't There*, Pan Macmillan, Introduction, para. 4.
[115] Duggan, M. (2020) Rhine Research Center, *Psi Encyclopedia*, https://psi-encyclopedia.spr.ac.uk/articles/rhine-research-center#footnote5_6iusr8h
[116] Nelson, R. (2018) Robert Jahn, *Psi Encyclopedia*, https://psi-encyclopedia.spr.ac.uk/articles/robert-jahn
[117] Radin, D. & Nelson, R. (n.d.) Meta-analysis of mind–matter interaction experiments: 1959 to 2000, *ResearchGate*, retrieved 5 September 2025, from https://www.researchgate.net/publication/241676080_Meta-analysis_of_mind-matter_interaction_experiments_1959_to_2000
[118] Nelson, R. (2019) *Connected: The Emergence of Global Consciousness*, ICRL Press.
[119] *Global Consciousness* (n.d.). Retrieved 9 July 2025, from https://gcp2.net/data-results/event-analysis
[120] *GCP Results* (n.d.) Retrieved 9 July 2025, from https://noosphere.princeton.edu/results.html
[121] Nelson, 2019.
[122] Holmberg, U. (2023) A novel market sentiment measure: Assessing the link between VIX and the Global Consciousness Projects data, *Journal of Economic Studies*, 51 (7), 1393–1409.
[123] Basios, V. & Janu, W. (2022, September 2) Nonlocal consciousness correlates analysis: The 'oREGano' project, *The Galileo Commission*, https://galileocommission.org/nonlocal-consciousness-correlates-analysis-the-oregano-project-vasileios-basios-wolfhardt-janu/
[124] Unpublished work—personal correspondence and unpublished presentations at Science and Consciousness conferences at Broughton Sanctuary, nr. Skipton, Yorkshire, UK.
[125] Wyrd Technologies (n.d.) *Wyrd Experience: How Wyrd Are You?*, retrieved 10 July 2025, from https://gowyrd.org/
[126] French, 2024, p. 13.
[127] Leibovici, L. (2001) Effects of remote, retroactive intercessory prayer on outcomes in patients with bloodstream infection: Randomised controlled trial, *British Medical Journal*, 323, 1450–1451.
[128] Lee, R.T., Kingstone, T., Roberts, L., Edwards, S., Soundy, A., Shah, P.R., Haque, M.S. & Singh, S. (2017) A pragmatic randomised controlled trial of healing therapy in a gastroenterology outpatient setting, *European Journal of Integrative Medicine*, 9, 110–119. https://doi.org/10.1016/j.eujim.2016.11.017, p. 118.
[129] Yu, C.-T.R., Radin, D.I., Chu, C.-Y. & Shiah, Y.-J. (2025) Effects of intentionally-treated water on cell migration of human glioblastoma cells, *EXPLORE*.

[130] Wikipedia (2025) Extraordinary claims require extraordinary evidence, *Wikipedia*, retrieved from https://en.wikipedia.org/w/index.php?title=Extraordinary_claims_require_extraordinary_evidence&oldid=1300219273

[131] Roe, 2023.

[132] Wahbeh, H., Radin, D., Mossbridge, J., Vieten, C. & Delorme, A. (2018) Exceptional experiences reported by scientists and engineers, *EXPLORE*, 14 (5), 329–341. https://doi.org/10.1016/j.explore.2018.05.002

[133] Cardeña, E. (2018) The experimental evidence for parapsychological phenomena: A review, *American Psychologist*, 73 (5), 663–677. https://doi.org/10.1037/amp0000236

[134] Funder, D.C. & Ozer, D.J. (2019) Evaluating effect size in psychological research: Sense and nonsense, *Advances in Methods and Practices in Psychological Science*, 2 (2), 156–168. https://doi.org/10.1177/2515245919847202

[135] Funder and Ozer, 2019.

[136] Butzer, B. (2020) Bias in the evaluation of psychology studies: A comparison of parapsychology versus neuroscience, *EXPLORE*, 16 (6), 382–391.

[137] Wikipedia (2025) List of topics characterized as pseudoscience, *Wikipedia*, retrieved from https://en.wikipedia.org/w/index.php?title=List_of_topics_characterized_as_pseudoscience&oldid=1306550596#Physical_sciences

[138] An interview with the founder of Guerilla Skeptics: Radford, B. (2022) The battle over Wikipedia misinformation: Interview with Guerilla Skeptic guru Susan Gerbic, *Center for Inquiry*, https://centerforinquiry.org/blog/the-battle-over-wikipedia-misinformation-interview-with-guerilla-skeptic-guru-susan-gerbic/

[139] For example by Rupert Sheldrake, one of the parapsychologists whose *Wikipedia* pages have been targeted by Guerilla Skeptics: Sheldrake, R. (n.d.) Wikipedia hijacked by dogmatists, *Rupert Sheldrake – Author and Biologist*, retrieved 5 January 2026, from https://www.sheldrake.org/reactions/wikipedia

[140] French, C.C. (2018) Reflections on pseudoscience and parapsychology: From here to there and (slightly) back again, in Pseudoscience: The conspiracy against science (pp. 375–391), *Boston Review*. https://doi.org/10.7551/mitpress/9780262037426.003.0017

[141] Cardeña, 2018.

[142] Reber, A.S. & Alcock, J.E. (2020) Searching for the impossible: Parapsychology's elusive quest, *American Psychologist*, 75 (3), 391–399. https://doi.org/10.1037/amp0000486

[143] Reber, A.S. & Alcock, J.E. (2019) Why parapsychological claims cannot be true, *Skeptical Inquirer*, https://skepticalinquirer.org/2019/07/why-parapsychological-claims-cannot-be-true/

[144] Sapolsky, R.M. (2023) *Determined: Life Without Free Will*, Random House, Chapter 1, para. 7.

[145] Hossenfelder, S. (2020) You don't have free will, but don't worry, *YouTube*, https://youtu.be/zpU_e3jh_FY?si=IQb2RMP5ZcgN7yQR

---

[146] Harris, S. (2012) Sam Harris on 'free will', *YouTube*, https://www.youtube.com/watch?v=pCofmZlC72g

[147] Al-Khalili, J. (2019) Do we have free will—A physicist's perspective? *Jim Al-Khalili Blog*, https://jimal-khalili.com/blog/do-we-have-free-will-a-physicists-perspective/

[148] Bowen, D. (2018) Daniel Dennett on Free Will, in *Encyclopedia of Evolutionary Psychological Science*, pp. 1–6, Springer, Cham. https://doi.org/10.1007/978-3-319-16999-6_2168-1

[149] Mozes, E. (2010).The dogmatic determinism of Daniel Dennett, *The Atlas Society*, https://www.atlassociety.org/post/the-dogmatic-determinism-of-daniel-dennett

[150] Quoted in Seth, A. & Snibbe, S. (2023) Free Will, consciousness, & reality with Dr. Anil Seth, *A Skeptic's Path to Enlightenment*, https://www.skepticspath.org/podcast/free-will-consciousness-reality-with-dr-anil-seth/

[151] Libet, B., Gleason, C.A., Wright, E.W. & Pearl, D.K. (1983) Time of conscious intention to act in relation to onset of cerebral activity (readiness-potential) the unconscious initiation of a freely voluntary act, *Brain*, 106 (3), 623–642.

[152] Harris, 2012.

[153] Parisetti, P.C., personal communication, 21 September 2025.

[154] Herrmann, C.S., Pauen, M., Min, B.K., Busch, N.A. & Rieger, J.W. (2008) Analysis of a choice-reaction task yields a new interpretation of Libet's experiments, *International Journal of Psychophysiology*, 67 (2), 151–157. https://doi.org/10.1016/j.ijpsycho.2007.10.013

[155] Schurger, A., Sitt, J.D. & Dehaene, S. (2012) An accumulator model for spontaneous neural activity prior to self-initiated movement, *Proceedings of the National Academy of Sciences of the United States of America*, 109 (42). https://doi.org/10.1073/pnas.1210467109

[156] Hermann et al., 2008, p. 151.

[157] Hossenfelder, 2020.

[158] Kastner, R.E. (2015) *Understanding Our Unseen Reality: Solving Quantum Riddles*, Imperial College Press, p. 183.

[159] Ibid., pp. 185–186.

[160] Harris, 2012.

[161] Sapolsky, 2023, paras. 25–26.

[162] Quoted in Spitz, H.H. (2004) Contemporary challenges to William James's white crow, *Skeptical Inquirer*, Jan/Feb 2004. https://cdn.centerforinquiry.org/wp-content/uploads/sites/29/2004/01/22164701/p51.pdf

[163] Zwiebach, B. (2016) *Key Features of Quantum Mechanics*.

[164] Ibid.

[165] Letter from Einstein to Max Born, 3 March 1947; The Born-Einstein Letters; Correspondence between Albert Einstein and Max and Hedwig Born from 1916 to 1955, Walker, New York, 1971. Cited in Hobson, M.P., et al. (1998) Quantum entanglement and communication complexity, *SIAM J. Comput.*, 30 (6): 1829–1841. CiteSeerX 10.1.1.20.8324

[166] Kotler, S., Peterson, G.A., Shojaee, E., Lecocq, F., Cicak, K., Kwiatkowski, A., Geller, S., Glancy, S., Knill, E., Simmonds, R.W., Aumentado, J. & Teufel, J.D. (2021) Direct observation of deterministic macroscopic entanglement, *Science*, 372 (6542), 622–625. https://doi.org/10.1126/science.abf2998

[167] Kristiansen, A. (2019) The future of quantum biology, *Royal Society Blog*, https://royalsociety.org/blog/2019/02/the-future-of-quantum-biology/

[168] Hameroff, S. & Penrose, R. (2014) Consciousness in the universe: A review of the 'Orch OR' theory, *Physics of Life Reviews*, 11 (1), 39–78. https://doi.org/10.1016/j.plrev.2013.08.002

[169] Penrose, R. (1990) *The Emperor's New Mind*, Vintage.

[170] Paulson, S. (2017) Roger Penrose on why consciousness does not compute, *Nautilus*, https://nautil.us/roger-penrose-on-why-consciousness-does-not-compute-236591/

[171] Sferra, A., Nicita, F. & Bertini, E. (2020. Microtubule dysfunction: A common feature of neurodegenerative diseases, *International Journal of Molecular Sciences*, 21 (19), 7354. https://doi.org/10.3390/ijms21197354

[172] Paulson, 2017.

[173] Anderson, J. & Hameroff, S. (2025) This scientist is revealing our QUANTUM consciousness [Video recording], *Project Unity*. https://rumble.com/v6mi1uc-this-scientist-is-revealing-our-quantum-consciousness.html

[174] Ibid.

[175] Adamatzky, A. (2012) Slime mould computes planar shapes, *International Journal of Bio-Inspired Computation*, 4 (3), 149–154.

[176] Nakagaki, T., Yamada, H. & Tóth, Á. (2000) Maze-solving by an amoeboid organism, *Nature*, 407 (6803), 470–470.

[177] Hameroff, S. & Penrose, R. (2014), p. 70.

[178] Stapp, H. (2015) A quantum-mechanical theory of the mind/brain connection, in E.F. Kelly, A. Crabtree & P. Marshall (eds.) *Beyond Physicalism: Toward Reconciliation of Science and Spirituality*, Rowman & Littlefield.

[179] Sivasundaram, S. & Nielsen, K.H. (2016) Surveying the attitudes of physicists concerning foundational issues of quantum mechanics, *arXivi*. http://arxiv.org/abs/1612.00676

[180] Faye, J. (2024) Copenhagen interpretation of quantum mechanics, *The Stanford Encyclopedia of Philosophy*. https://plato.stanford.edu/archives/sum2024/entries/qm-copenhagen

[181] Bacciagaluppi, G. (2025) The role of decoherence in quantum mechanics, *Stanford Encyclopedia of Philosophy*. https://plato.stanford.edu/entries/qm-decoherence/

[182] Smith Churchland, P. (1997) Brainshy: Non-neural theories of conscious experience, *Consciousness – Papers for Tucson II*. https://patriciachurchland.com/wp-content/uploads/2020/07/1997-Brainshy-NonNeural-Theories-of-Conscious-Experience.pdf

[183] Robinson, H. & Weir, R. (2025) Dualism, in E.N. Zalta & U. Nodelman (eds.) *The Stanford Encyclopedia of Philosophy (Winter 2025)*. https://plato.stanford.edu/archives/win2025/entries/dualism/

[184] Hameroff & Penrose, 2014.

185 Ibid., p. 66.
186 Anderson & Hameroff, 2025.
187 Quotes in this section are all taken from Jahn, R.G. & Dunne, B.J. (2012) *Quirks of the Quantum Mind*, ICRL Press, Appendix B.
188 Hameroff, S. & Chopra, D. (2012) The 'quantum soul': A scientific hypothesis, in *Exploring Frontiers of the Mind–Brain Relationship*, Springer.
189 Ibid., p. 89.
190 Anderson & Hameroff, 2025.
191 Ibid.
192 Stapp, 2015.
193 Anderson & Hameroff, 2025.
194 Ibid.
195 Hameroff & Penrose, 2014, p. 64, emphasis original.
196 Angulo, D., Thompson, K., Nixon, V.-M., Jiao, A., Wiseman, H.M. & Steinberg, A.M. (2024) Experimental evidence that a photon can spend a negative amount of time in an atom cloud (No. arXiv:2409.03680), *arXiv*. https://doi.org/10.48550/arXiv.2409.03680
197 Stapp, 2015.
198 Hameroff, S. (2012) How quantum brain biology can rescue conscious free will, *Frontiers in Integrative Neuroscience*, 6 (SEPTEMBER), 1–17. https://doi.org/10.3389/fnint.2012.00093
199 Stapp, 2015.
200 Gardner, M. (2003) *Relativity Simply Explained*, Dover Publications Inc.
201 Cheng, T.-P. (2013) The Kaluza–Klein theory and extra dimensions, in T.-P. Cheng (Ed.) *Einstein's Physics: Atoms, Quanta, and Relativity – Derived, Explained, and Appraised*, Oxford University Press. https://doi.org/10.1093/acprof:oso/9780199669912.003.0017
202 Greene, B. (1999) *The Elegant Universe – Superstrings, Hidden Dimensions & the Quest for the Ultimate Theory*, W.W. Norton and Co.
203 M-theory, the theory formerly known as Strings, *The Stephen Hawking Centre for Theoretical Cosmology*. Retrieved 10 August 2024 from https://www.ctc.cam.ac.uk/outreach/origins/quantum_cosmology_four.php
204 Sutter, P. (2020) How the universe could possibly have more dimensions, *Space.com*. https://www.space.com/more-universe-dimensions-for-string-theory.html
205 Carr, B. (2010) A proposed new paradigm of matter, mind and spirit, pt 1, *Scientific & Medical Network Review*, Spring, 3–8.
206 Carr, B. (2010) A proposed new paradigm of matter, mind and spirit, pt 2, *Scientific & Medical Network Review*, Summer, 3–8.
207 Carr, B. (2015) Hyperspatial models of matter and mind, in E.F. Kelly (Ed.) *Beyond Physicalism: Toward Reconciliation of Science and Spirituality*, Rowman & Littlefield.
208 Carr, B. (2021) Making space and time for consciousness in physics, in P. Dennison (Ed.) *Perspectives on Consciousness* (pp. 319–350), Nova Science Publishers.

---

[209] Peterson, R. (1997) *Out of Body Experiences: How to Have Them and What to Expect*, Hampton Roads Publishing Company, Inc.

[210] Kadagian, D.J., Shushan, G. & Van Lomell, P. (2022) *The Crossover Experience: Life After Death / 100 Exceptional Near Death Experiences*, Four Seasons Productions LLC.

[211] Abbott, E.A. (1884) *Flatland: A Romance of Many Dimensions*, Seeley & Co.

[212] Carr, 2015, *The Universal Structure…*, para. 5.

[213] Clay, E.R. (1882) *The Alternative: A Study in Psychology*, Oxford University Press.

[214] Neppe, V.M. & Close, E.R. (2014) Reality begins with consciousness: A paradigm shift that works, *Brain Voyage*.

[215] Neppe, V.M. & Close, E.R. (2020) The Neppe-Close triadic dimensional vortical paradigm: An invited summary, *International Journal of Physics Research and Applications*, 3, 001–014. https://doi.org/10.29328/journal.ijpra.1001018

[216] Neppe & Close, 2014, p. 93.

[217] Kabat-Zinn, J. (2025) *Jon Kabat-Zinn*. https://jonkabat-zinn.com/

[218] Tolle, E. (2001) *The Power of Now: A Guide to Spiritual Enlightenment*, Yogi Impressions Books Pvt Ltd.

[219] Neppe & Close, 2014, p. 12.

[220] Raskin, R.A.L. (n.d.) Gimmel—The third letter of the Hebrew alphabet, *Chabad.org*. Retrieved 23 April 2025, from https://www.chabad.org/library/article_cdo/aid/137075/jewish/Gimmel.htm

[221] Neppe & Close, 2020.

[222] Neppe & Close, 2014, p. 13.

[223] Neppe & Close, 2020.

[224] New Thinking Allowed with Jeffrey Mishlove (Director) (2020, August 10) *Psychical Research and Parapsychology with Bernard Carr* [Video recording]. https://www.youtube.com/watch?v=_MYo74u97GM

[225] Sharma, P. & Tucker, J.B. (2004) Cases of reincarnation type with memories from the intermission between lives, *Journal of Near-Death Studies*, 23(2), 101–118. https://doi.org/10.17514/jnds-2004-23-2-p101-118

[226] Carr, 2015, para. 1.

[227] Angulo, D., et al., 2024.

[228] Grosso, M. (2015) The 'transmission' model of mind and body, in E.F. Kelly, A. Crabtree & P. Marshall (Eds.) *Beyond Physicalism*, Rowman & Littlefield.

[229] Neppe & Close, 2014, p. 102.

[230] Ibid., p. xxxii.

[231] Alfonseca, M. (2016) Free-will in the block universe? Old and new problems with time travels, *Philosophia*.

[232] Jones, M.W. & Hunt, T. (2023) Electromagnetic-field theories of qualia: Can they improve upon standard neuroscience? *Frontiers in Psychology*, 14 (June), 1–18. https://doi.org/10.3389/fpsyg.2023.1015967

[233] Hennacy, D.P. (2015) Autistics, savants, and psi: A radical theory of mind, *EdgeScience*, 23, 12–18.

[234] Hennacy, D.M. (2024) The neuroscience of savant syndrome, enlightenment, and other extraordinary states, in L.J. Miller (Ed.) *The Oxford Handbook of Psychology and Spirituality*, Oxford University Press. https://doi.org/10.1093/oxfordhb/9780190905538.013.33
[235] Koob, A. (2009) *The Root of Thought: Unlocking Glia – the Brain Cell that Will Help Us Sharpen Our Wits, Heal Injury, and Treat Brain Disease*, FT Press.
[236] Størmer, F.C. (2014) Is magnetite a universal memory molecule? *Medical Hypotheses*, 83 (5), 549–551. https://doi.org/10.1016/j.mehy.2014.08.028
[237] Hennacy, 2024, Neuromelanin…, para. 4.
[238] Lokhorst, G.-J. (2021. Descartes and the pineal gland, in E.N. Zalta (Ed.) *The Stanford Encyclopedia of Philosophy (Winter 2021)*. https://plato.stanford.edu/archives/win2021/entries/pineal-gland/
[239] Roney-Dougal, S. (1989) Recent findings relating to the possible role of the pineal gland in affecting psychic ability, *Journal of the Society for Psychical Research*, **55** (815), 313–328.
[240] Swanson, C. (2008) The torsion field and the aura, *Subtle Energies Energy Medicine*, 19 (3), 43–89.
[241] Sinclair, U. (1930) *Mental Radio*, Charles C Thomas Publishers.
[242] Sheldrake, R. (n.d.) Morphic resonance and morphic fields—an introduction, *Sheldrake.Org*. Retrieved 8 December 2025, from https://www.sheldrake.org/research/morphic-resonance/introduction
[243] Sheldrake, R. (1995) *A New Science of Life: The Hypothesis of Morphic Resonance*, Park Street Press.
[244] For example, see Arnold, D. (2022) Where are memories stored in the brain? New research suggests they may be in the connections between your brain cells, *The Conversation*. http://theconversation.com/where-are-memories-stored-in-the-brain-new-research-suggests-they-may-be-in-the-connections-between-your-brain-cells-174578
[245] Blackiston, D.J., Silva Casey, E. & Weiss, M.R. (2008) Retention of memory through metamorphosis: Can a moth remember what it learned as a caterpillar? *PLoS ONE*, 3 (3), e1736. https://doi.org/10.1371/journal.pone.0001736
[246] Sheldrake, R. (1987) Mind, memory, and archetype morphic resonance and the collective unconscious, *Psychological Perspectives*, 18 (1), 9–25.
[247] Editors (2025) Gene, *Encyclopaedia Britannica*. https://www.britannica.com/science/gene
[248] Sheldrake, 1987.
[249] Lemetyinen, H. (2023, September 7) Language acquisition theory in psychology, *Simply Psychology*. https://www.simplypsychology.org/language.html
[250] Sheldrake, R. (2006) Morphic fields, *World Futures*, 62 (1–2), 31–41. https://doi.org/10.1080/02604020500406248
[251] Sheldrake, 1987.
[252] Archer, R. (2025) The telepathy tapes—Dr. Diane Hennacy Powell—Buddha at the gas pump interview, *YouTube*, https://www.youtube.com/watch?v=qECSibSj3dU

[253] Keating, S. (2020) The violent attack that turned a man into a maths genius, *BBC Online*. https://www.bbc.com/future/article/20190411-the-violent-attack-that-turned-a-man-into-a-maths-genius

[254] Brogaard, B. (2012) Kim Peek, the real rain man, *Psychology Today*. https://www.psychologytoday.com/us/blog/the-superhuman-mind/201212/kim-peek-the-real-rain-man

[255] Hennacy, 2024.

[256] Sheldrake, R. (2019) What happens when we die? *YouTube*. https://www.youtube.com/watch?v=vWuR7Z9AcH0

[257] Editors (20025) Life after death: Near-death experiences of hindus, *Sanskriti – Hinduism and Indian Culture Website*. https://www.sanskritimagazine.com/life-death-near-death-experiences-hindus/

[258] Sheldrake, 2019.

[259] Hennacy, 2015.

[260] Storm, L. & Tressoldi, P.E. (2020) Meta-analysis of free-response studies 2009-2018: Assessing the noise-reduction model ten years on, *Journal of the Society for Psychical Research*, 84 (4), 193–219.

[261] Baker, B. (2013) Deep secret—secure submarine communication on a quantum level, *Naval Technology*. https://www.naval-technology.com/features/featuredeep-secret-secure-submarine-communication-on-a-quantum-level/?cf-view&cf-closed

[262] Boyle, L. (2024) On interstellar quantum communication and the Fermi paradox (No. arXiv:2408.02445), *arXiv*. https://doi.org/10.48550/arXiv.2408.02445

[263] Sheldrake, R. (2003) *The Sense of Being Stared At*, Random House.

[264] Archer, 2025.

[265] Before Skool (2025) Do humans have free will? - Rupert Sheldrake [Video recording], *YouTube*. https://www.youtube.com/watch?v=esoSLmnYpGQ

[266] Fallon, F. (n.d.) Integrated information theory of consciousness, *Internet Encyclopedia of Philosophy*. https://iep.utm.edu/integrated-information-theory-of-consciousness/

[267] Tononi, G. (2008) Consciousness as integrated information: A provisional manifesto, *The Biological Bulletin*, 215 (3), 216–242. https://doi.org/10.2307/25470707

[268] Ibid.

[269] Ibid.

[270] Albantakis, L., Barbosa, L., Findlay, G., Grasso, M., Haun, A.M., Marshall, W., Mayner, W.G.P., Zaeemzadeh, A., Boly, M., Juel, B.E., Sasai, S., Fujii, K., David, I., Hendren, J., Lang, J.P. & Tononi, G. (2023) Integrated information theory (IIT) 4.0: Formulating the properties of phenomenal existence in physical terms, *PLoS Computational Biology*, 19 (10), e1011465. https://doi.org/10.1371/journal.pcbi.1011465

[271] Ibid.

[272] Ibid.

[273] Ibid.

[274] Tononi, 2008, p. 232.

[275] Ibid, p. 233.
[276] Ibid.
[277] Kastrup, B. (2023) In defense of integrated information theory (IIT), *Essentia Foundation*. https://www.essentiafoundation.org/in-defense-of-integrated-information-theory-iit/reading/
[278] Cea, I., Negro, N. & Signorelli, C.M. (2023) The fundamental tension in integrated information theory 4.0's realist idealism, *Entropy*, 25 (10), 1453, p. 1. https://doi.org/10.3390/e25101453
[279] Ibid., p. 8.
[280] The summary of Campbell's theory in this section is derived from throughout his trilogy that was published by Lightning Strike Books in 2003 and comprises: *My Big TOE: Awakening: Book 1 of a Trilogy Unifying Philosophy, Physics and Metaphysics*; *My Big TOE: Discovery: Book 2*; and *My Big TOE: Inner Workings: Book 3*. Direct quotes from these volumes will be specifically referenced but otherwise the source for the concepts described is not specific to any of these books in particular.
[281] Campbell, Book 1, 2003, p. 61.
[282] Monroe, R.A. (1972) *Journeys Out of the Body*, Souvenir Press Ltd.
[283] This led to forming the Monroe Institute, which still offers programmes related to OBEs: https://www.monroeinstitute.org/
[284] Campbell, Book 1, 2003, p. 94.
[285] Ibid., p. 102.
[286] Campbell, Book 3, 2003, p. 198.
[287] Campbell, Book 1, 2003, p. 274.
[288] Campbell, Book 3, 2003, p. 78.
[289] Campbell, Book 2, 2003, p. 293.
[290] Campbell, Book 1, 2003, p. 309.
[291] Ibid., p. 336.
[292] Campbell, Book 3, 2003, p. 194.
[293] Campbell, Book 1, 2003, p. 204.
[294] Campbell, Book 2, 2003, p. 102.
[295] Fallon, n.d.
[296] Campbell, Book 2, 2003, p. 104.
[297] Tononi, G. (2012) *Phi: A Voyage from the Brain to the Soul* (Kindle), Knopf Doubleday Publishing Group, p. 250.
[298] Ibid., p. 4.
[299] Ibid., p. 11.
[300] Ibid., p. 238.
[301] Campbell, Book 3, 2003, p. 140.
[302] Campbell, Book 2, 2003, p. 300.
[303] Ibid., p. 145.
[304] Ibid., p. 250.
[305] Tononi, 2012, p. 21.
[306] Campbell, Book 2, 2003, p. 179.
[307] Ibid., p. 299.

[308] Greyson, B. (1983) The near-death experience scale: Construction, reliability, and validity, *The Journal of Nervous and Mental Disease*, 171 (6), 369–375.

[309] Tononi, G., Albantakis, L., Boly, M., Cirelli, C. & Koch, C. (2022) Only what exists can cause: An intrinsic view of free will, *arXiv*, preprint, arXiv:2206.02069, 1–26, p. 1.

[310] Campbell, Book 2, 2003, p. 370.

[311] The material in this chapter is summarised from two books, both published by BenBella Books: Lanza, R. & Berman, B. (2010) *Biocentrism: How Life and Consciousness are the Keys to Understanding the True Nature of the Universe*; and Lanza, R., Pavsic, M. & Berman, B. (2020) *The Grand Biocentric Design: How Life Creates Reality*.

[312] Lanza & Berman, 2010, pp. 112–113.

[313] Kovachy, T., Asenbaum, P., Overstreet, C., Donnelly, C.A., Dickerson, S.M., Sugarbaker, A., Hogan, J.M. & Kasevich, M.A. (2015) Quantum superposition at the half-metre scale, *Nature*, 528 (7583), 530–533. https://doi.org/10.1038/nature16155

[314] Marletto, C., Coles, D.M., Farrow, T. & Vedral, V. (2018) Entanglement between living bacteria and quantized light witnessed by Rabi splitting, *Journal of Physics Communications*, 2 (10), 101001. https://doi.org/10.1088/2399-6528/aae224

[315] Proietti, M., Pickston, A., Graffitti, F., Barrow, P., Kundys, D., Branciard, C., Ringbauer, M. & Fedrizzi, A. (2019) Experimental test of local observer independence, *Science Advances*, 5 (9), eaaw9832. https://doi.org/10.1126/sciadv.aaw9832

[316] An excellent overview of this question is provided in Chapter 2 of Goff, P. (2023) *Why? The Purpose of the Universe*, OUP Oxford.

[317] Lanza & Berman, 2010, p. 90.

[318] Hoffman, D.D. (2024) Plenary 3—TSC—'consciousness and reality', *YouTube*.
https://www.youtube.com/watch?v=_YK4ML3HLgo&list=LL&index=1&t=2501s

[319] Hoffman, D.D., Prakash, C. & Prentner, R. (2023) Fusions of consciousness, *Entropy*, 25 (1), 1–40. https://doi.org/10.3390/e25010129

[320] Wachowski, L. & Wachowski, L. (Directors) (1999) *The Matrix* [Science fiction].

[321] Hoffman, D.D. (2020) *The Case Against Reality: How Evolution Hid the Truth from Our Eyes*, Penguin.

[322] Hoffman, D.D., Singh, M. & Prakash, C. (2015) The Interface Theory of Perception, *Psychonomic Bulletin and Review*, 22 (6), 1480–1506. https://doi.org/10.3758/s13423-015-0890-8

[323] Prakash, C., Stephens, K.D., Hoffman, D.D., Singh, M. & Fields, C. (2021) Fitness beats truth in the evolution of perception, *Acta Biotheoretica*, 69 (3), 319–341. https://doi.org/10.1007/s10441-020-09400-0

[324] Hoffman, 2021.

[325] Hoffman, 2020, quoted on p. 116.

---

326 Hoffman, 2024.
327 UNIVERSE+ (2024) Positive geometry in particle physics and cosmology! *Positive Geometry in Particle Physics and Cosmology*. https://positive-geometry.com/index.shtml
328 Hoffman, 2024.
329 Ibid.
330 Lanza, Pavsic & Berman, 2020, p. 126.
331 Hoffman, 2020, p. 201.
332 Lanza et al., 2020, p. 72.
333 Ibid., p. 37.
334 Ibid., p. 37.
335 Know Thyself (2023) Understanding the illusion of free will—with Donald Hoffman, *YouTube*. https://youtu.be/oR0l4ihSF3o?si=7x27XN2z52cYD_Am
336 Kuhn, 2024.
337 Hoffman, 2024.
338 Kastrup, 2023.
339 Dawkins, R. (1976) *The Selfish Gene*, Oxford University Press, Preface, para. 1.
340 For example, see Kirk, R. (2005) *Zombies and Consciousness*, Oxford University Press.
341 Allen, C. & Trestman, M. (2024) Animal consciousness, *The Stanford Encyclopedia of Philosophy*. https://plato.sydney.edu.au/entries/consciousness-animal/
342 Nieder, A., Wagener, L. & Rinnert, P. (2020) A neural correlate of sensory consciousness in a corvid bird, *Science*, 369 (6511), 1626–1629. https://doi.org/10.1126/science.abb1447
343 Billard, P., Clayton, N.S. & Jozet-Alves, C. (2020) Cuttlefish retrieve whether they smelt or saw a previously encountered item, *Scientific Reports*, 10 (1), 5413. https://doi.org/10.1038/s41598-020-62335-x
344 Galpayage Dona, H.S., Solvi, C., Kowalewska, A., Mäkelä, K., MaBouDi, H. & Chittka, L. (2022) Do bumble bees play? *Animal Behaviour*, 194, 239–251. https://doi.org/10.1016/j.anbehav.2022.08.013
345 Low, P. (2012) The Cambridge Declaration on Consciousness, *Proceedings of the Francis Crick Memorial Conference*, Churchill College, Cambridge University.
346 Andrews, K., Birch, J., Sebo, J. & Sims, T. (2024) The New York Declaration on Animal Consciousness, *Background to the New York Declaration on Animal Consciousness*. nydeclaration.com
347 Segundo-Ortin, M. & Calvo, P. (2022) Consciousness and cognition in plants, *Wiley Interdisciplinary Reviews: Cognitive Science*, 13 (2), e1578.
348 Horgan, J. (2023) A 25-year-old bet about consciousness has finally been settled, *Scientific American*. https://www.scientificamerican.com/article/a-25-year-old-bet-about-consciousness-has-finally-been-settled/
349 Frigg, R. (2006) Scientific representation and the semantic view of theories, *Theoria*, 55 (2), 183–206.

350 Roe, C.A. (2019) The egregious state of scepticism, *Paranormal Review*, 90, 4–5.

351 Popper, K.R. (1959) *The Logic of Scientific Discovery*, Basic Books.

352 Dawkins, R. (2006) *The God Delusion*, Bantam Press.

353 ConTraSt Database: https://contrastdb.tau.ac.il/

354 Yaron, I., Melloni, L., Pitts, M. & Mudrik, L. (2022) The ConTraSt database for analysing and comparing empirical studies of consciousness theories, *Nature Human Behaviour*, 6 (4), 593–604. https://doi.org/10.1038/s41562-021-01284-5

355 Ibid.

356 Del Pin, S.H., Skóra, Z., Sandberg, K., Overgaard, M. & Wierzchoń, M. (2021) Comparing theories of consciousness: Why it matters and how to do it, *Neuroscience of Consciousness*, 2021 (2), 1–8. https://doi.org/10.1093/nc/niab019

357 This is somewhat similar to Gertrude Schmeidler's suggestion of there being two distinct stages to the processing of psychic information. The first stage concerns how psychic information enters the organism, a question that is still very much unresolved but which should be accommodated by good theories of non-local consciousness. The second stage refers to how this information is processed thereafter to produce psychic experiences, and can be addressed using mainstream psychology and neuroscience. Schmeidler, G. (1986) Subliminal perception and ESP: Order in diversity? *Journal of the American Society for Psychical Research*, 80, 241–264.

358 Nahm, M. (2022) Toward a "parapsychological synthesis": Proposals for integrating theories of psi, *Abstracts of the Combined Conference of the Parapsychological Association and the Society for Scientific Exploration*, 240–242. https://doi.org/10.30891/jopar.2022.02.03

359 Kirchhoff, M.D. & Kiverstein, J. (2019) *Extended Consciousness and Predictive Processing: A Third Wave View*, Routledge. https://doi.org/10.4324/9781315150420

360 Grosso, 2015.

361 Quoted in Preface, para. 8 of Luke, D. (2019) *Otherworlds: Psychedelics and Exceptional Human Experience*, Aeon Books.

362 Freedman, M., Binns, M.A., Meltzer, J.A., Hashimi, R. & Chen, R. (2024) Enhanced mind-matter interactions following rTMS induced frontal lobe inhibition, *Cortex*, 172, 222–233. https://doi.org/10.1016/j.cortex.2023.10.016

363 Ibid.

364 Bancel, P.A. (2017) Searching for global consciousness: A 17-year exploration, *Explore: The Journal of Science and Healing*, 13 (2), 94–101. https://doi.org/10.1016/j.explore.2016.12.003

365 Broughton, R. (2024) Fifty years among the RNGs, and why we know they are not consciousness detectors, *Book of Abstracts: 47th SPR International Annual Conference*, Derby, UK. https://www.spr.ac.uk/sites/default/files/civicrm/persist/contribute/files/SPR%202024%20Book%20of%20Abstracts%20with%20Tuijl.pdf

---

366 Duggan, M. (2020) Richard S Broughton, *Psi Encyclopedia*. https://psi-encyclopedia.spr.ac.uk/articles/richard-s-broughton#Improving_Research_Methodology
367 Editors (2024) About us, *Global Consciousness Project 2.0*. https://gcp2.net/about. Retrieved 3 December 2025.
368 Radin, D., personal communication, 7 November 2025.
369 Kauffman, S. & Radin, D. (2021) Is the brain-mind quantum? A theoretical proposal with supporting evidence, *arXiv*, preprint arXiv:2101.01538, 1–25.
370 Radin, D., Michel, L., Galdamez, K., Wendland, P., Rickenbach, R. & Delorme, A. (2012) Consciousness and the double-slit interference pattern: Six experiments, *Physics Essays*, 25 (2), 157–171. https://doi.org/10.4006/0836-1398-25.2.157
371 Radin, D., Wahbeh, H., Michel, L. & Delorme, A. (2019) Psychophysical effects in double-slit interference patterns: Response to a critique, *PsyArXiv*. https://doi.org/10.31234/osf.io/9csgu
372 Radin et al., 2012, p. 157.
373 Kastner, 2015.
374 Everett, H. (1957) A many-worlds formulation of quantum theory, *Reviews of Modern Physics*, 29, 454.
375 Kauffman and Radin, 2021, p. 4.
376 Institute of Noetic Sciences. *Do you have psi abilities?* Retrieved 18 September 2024, from https://psiarcade.org/
377 Radin, D. (2023) What does precognition 'see'? *Society for Psychical Research Web Event*. https://www.spr.ac.uk/civicrm/event/info?id=193&reset=1
378 Steinkamp, F. (1997) Backwards causation, precognition and the intervention paradox, *The Parapsychological Association 40th Annual Convention, Proceedings of Presented Papers*, 410–425.
379 Parsons, S. (2022) Electromagnetism and paranormal phenomena, *Psi Encyclopedia*. https://psi-encyclopedia.spr.ac.uk/articles/electromagnetism-and-paranormal-phenomena
380 Schwartz, S.A. (1979) Two application-oriented experiments employing a submarine involving a novel remote viewing protocol, one testing the ELF hypothesis, *Proceedings American Society for Psychical Research*.
381 Ibid., p. 12.
382 Schwartz, S. (2014) Through time and space: The evidence for remote viewing, in D. Broderick & B. Groetzel (Eds.) *The Evidence for Psi: Essays on the Reality of Paranormal Phenomena: Thirteen Empirical Research Reports*, McFarland & Company, p. 31.
383 Boyle, 2024.
384 Sheldrake, R. (1994) *Seven Experiments that Could Change the World: A Do-It-Yourself Guide to Revolutionary Science*, Riverhead Books.
385 Ibid., p. 145.
386 Ibid., p. 262.
387 Vernon, D. & Roe, C.A. (2021) An implicit and explicit assessment of morphic resonance theory using Chinese characters, *Journal of the Society for Psychical Research*, 85 (3).